AF832168

Synthesis Lectures on Communications

This series of short books cover a wide array of topics, current issues, and advances in key areas of wireless, optical, and wired communications. The series also focuses on fundamentals and tutorial surveys to enhance an understanding of communication theory and applications for engineers.

Muhammad Ali Jamshed · Ali Nauman
Editors

Integrated Terrestrial and Non-Terrestrial Networks

Springer

Editors
Muhammad Ali Jamshed
College of Science and Engineering
University of Glasgow
Glasgow, UK

Ali Nauman
Information and Communication Engineering
Yeungnam University
Gyeongsan, Korea (Republic of)

ISSN 1932-1244 ISSN 1932-1708 (electronic)
Synthesis Lectures on Communications
ISBN 978-3-031-63704-9 ISBN 978-3-031-63705-6 (eBook)
https://doi.org/10.1007/978-3-031-63705-6

© The Editor(s) (if applicable) and The Author(s), under exclusive license to Springer Nature Switzerland AG 2025

This work is subject to copyright. All rights are solely and exclusively licensed by the Publisher, whether the whole or part of the material is concerned, specifically the rights of translation, reprinting, reuse of illustrations, recitation, broadcasting, reproduction on microfilms or in any other physical way, and transmission or information storage and retrieval, electronic adaptation, computer software, or by similar or dissimilar methodology now known or hereafter developed.
The use of general descriptive names, registered names, trademarks, service marks, etc. in this publication does not imply, even in the absence of a specific statement, that such names are exempt from the relevant protective laws and regulations and therefore free for general use.
The publisher, the authors and the editors are safe to assume that the advice and information in this book are believed to be true and accurate at the date of publication. Neither the publisher nor the authors or the editors give a warranty, expressed or implied, with respect to the material contained herein or for any errors or omissions that may have been made. The publisher remains neutral with regard to jurisdictional claims in published maps and institutional affiliations.

This Springer imprint is published by the registered company Springer Nature Switzerland AG
The registered company address is: Gewerbestrasse 11, 6330 Cham, Switzerland

If disposing of this product, please recycle the paper.

Contents

Integrated Terrestrial and Non-Terrestrial Network: An Overview

Bushra Haq, Muhammad Ali Jamshed, and Ali Nauman

Abstract

The fusion of terrestrial and non-terrestrial networks promises to bridge the connectivity gap for remote places and improve connectivity for those already online. This amalgamation heralds a significant shift in both technological impact and social progress, charting a course for long-term advancements. It marks a substantial collaborative challenge for the mobile and satellite communications sectors to design and manage a novel three-dimensional wireless network architecture. In its Release 17, the Third-generation partnership project (3GPP) has taken an important step towards this integration by standardizing non-terrestrial networks (NTN), enabling 5G terrestrial infrastructures supporting NTN framework. This framework is intended to include a network of low Earth orbit (LEO) satellites, with the inclusion of unmanned aerial vehicles (UAVs) and hhigh altitude platforms (HAPs) serving as distinct NTN use cases. However, like the global positioning system (GPS), NTN signals struggle to penetrate indoor environments due to attenuation and scattering by physical barriers such as roofs and walls. Furthermore,

B. Haq (✉)
Balochistan University of Information Technology, Engineering and Management Sciences Pakistan, Quetta, Pakistan
e-mail: bushra.haq@buitms.edu.pk

M. A. Jamshed
College of Science and Engineering, University of Glasgow, Glasgow G12 8QQ, UK

A. Nauman
Department of Information and Communication Engineering, Yeungnam University, Gyeongsan-si, Republic of Korea

© The Author(s), under exclusive license to Springer Nature Switzerland AG 2025
M. A. Jamshed and A. Nauman (eds.), *Integrated Terrestrial and Non-Terrestrial Networks*, Synthesis Lectures on Communications,
https://doi.org/10.1007/978-3-031-63705-6_1

the constraints of power and latency present considerable challenges to ensuring reliable,seamless coverage and low-latency communication for cellular devices utilizing NTN. This chapter provides an overview of NTNs and their integration into terrestrial networks (TN). We also provide information on various use cases of NTN and how the NTN can be integrated with 5G communication systems.

Keywords
Non-terrestrial networks (NTN) • Third-generation partnership project (3GPP) • Beyond 5G (B5G) • 6G

1 Introduction

Networks beyond Earth, termed as NTN, encompass UAVs, HAPs, and satellite networks. Traditionally, these networks have been utilized for specific purposes like disaster management, navigation, television broadcasting, and remote sensing. However, recent significant advancements in aerial and space technologies, coupled with reduced manufacturing and launching costs, have opened up avenues for more sophisticated applications of NTN. These advancements have been particularly noteworthy when these networks are integrated with terrestrial communication networks. This integration has spurred the envisioning of various novel use cases and applications, primarily focused on delivering continuous, pervasive, and high-capacity connectivity worldwide [1, 2].

The limitations inherent in ground-based infrastructure, alongside economic considerations, often hinder the deployment of TNs in remote or inaccessible regions such as rural areas, deserts, and oceans. Consequently, user equipments (UEs) remains unable to access terrestrial services within these underserved areas. Leveraging NTN alongside existing terrestrial infrastructure emerges as a viable and cost-effective solution to ensure uninterrupted and widespread wireless coverage. This collaboration holds the potential to facilitate network scalability and enable continuous connectivity across such areas [3, 4].

In instances like these, NTN serve as pivotal access points to enhance the performance of existing land-based networks concerning their capacity, reach, and speed. They offer a solution to the deficiencies of land-based infrastructure in meeting the necessary reliability standards and widespread availability required for upcoming wireless applications. Recognizing the advantages provided by NTN and aiming to leverage economies of scale, the 3GPP is exploring the integration of satellite access into the 5G ecosystem [1, 2]. Furthermore, as the 5G ecosystem develops and refines its specifications, especially in catering to Enhanced Mobile Broadband (eMBB) and Ultra-Reliable Low-Latency Communication (uRLLC), it enables a seamless inclusion of aerial vehicles within land-based networks. The capabilities of 5G technology meet the crucial demands of UAVs in terms of control and communication, enabling dependable connectivity for essential beyond visual line-of-sight operations [5–7], crucial for numerous applications. Emerging applications like autonomous vehicles, precision agriculture, and various undiscovered use cases are steering extensive research efforts towards 6G to establish global connectivity. The necessity for worldwide coverage, coupled with imminent plans for extensive LEO satellite constellations, will amplify the significance of NTN in future networks. Undoubtedly, fully integrated NTN can offer ubiquitous connectivity, delivering anything, anywhere, at any time, resulting in significant socio-economic impact [4]. Hence, it is imperative to provide a comprehensive overview of the integration of NTN into current and future wireless networks.

1.1 Motivation and Objectives

In this chapter, we discuss the basic working of NTN and their integration in TNs. The discussion encompasses the fundamental definition of NTN, deployment scenarios of NTN, and the integration of NTN in 5G communication. Specifically, we explore key challenges necessitating the integration of NTN in TN. Moreover, this chapter highlights the basic working mechanism of NTN and various use case scenarios making it a valuable resource for those working in this field.

1.2 Chapter Organization

The rest of the chapter is structured as follows: Sect. 2 presents the detailed definition of TN and NTN, this section also provides the various deployment configurations of NTN and the integrated architecture of TN & NTN. Following that, Sect. 3 outlines the NTN use cases, while Sect. 4 offers a detailed explanation of the integration of NTN in 5G systems. Finally, Sect. 5 concludes the chapter.

2 TN and NTN Definitions and Architecture

2.1 TN Definition

Terrestrial networks are the backbone of modern communication, spanning a broad array of technologies that have evolved dramatically over the past few decades. Initially, these networks were based on wired connections, like telephone lines and coaxial cables, which provided the main routes for voice and data communication. With the advent of cellular technology, wireless terrestrial networks have taken center stage, drastically transforming how we interact and access information.

The key characteristics of terrestrial networks include their reliance on physical infrastructure, such as cell towers, fiber-optic cables, and routing centers, strategically distributed across the landscape. This infrastructure supports multiple generations of cellular technologies, each marked by significant advancements in speed, capacity, and efficiency. From the analog signals of 1G to the high-speed digital communications of 4G, and now the emergence of 5G, terrestrial networks have continued to expand their capabilities.

The most recent, 5G, promises not just faster internet speeds but also lower latency, increased reliability, and the ability to connect many more devices simultaneously. This is particularly important for the internet of things (IoT), where countless devices require a stable and continuous internet connection. TNs have also adopted advanced technologies such as Multiple Input Multiple Output (MIMO) antennas, beam-forming, and small cells, which enhance signal quality and network capacity.

One of the critical strengths of terrestrial networks lies in their widespread coverage and established infrastructure, which makes them highly accessible to a large portion of the global population. Moreover, these networks are designed to be robust and secure, offering reliable communication channels for personal, commercial, and emergency services.

However, the terrestrial network is not without its limitations. The quality of coverage can vary significantly, with rural and remote areas often experiencing poorer service due to the economic and logistical challenges of extending infrastructure to these locations. Moreover, the physical nature of the infrastructure means it can be susceptible to damage from natural disasters, leading to potential service disruption. To address these challenges, there is a growing interest in integrating terrestrial networks with NTN.

2.2 NTN Definition

NTN reach far beyond the bounds of Earth's surface, comprising elements like satellites orbiting in space, UAVs soaring in our skies, and HAPs hovering in the stratosphere. Traditionally, these high-flying technologies have been deployed for specific roles. Satellites, for instance, have been instrumental in TV broadcasting, guiding navigation systems, and scanning the Earth for remote sensing data. UAVs and HAPs, on the other hand, often spring into action during emergencies, such as natural disasters, providing critical communication links where ground infrastructure is damaged or non-existent.

The deployment options for NTN can vary based on the specific type of platform utilized. These platforms come under two primary categorizations: airborne and spaceborne. Spaceborne are classified based on orbit, altitude, and beam footprint size. They can be categorized as follows:

- Geostationary Earth Orbiting (GEO), which maintains a fixed position in airspace for observing ground due to its circular, equatorial orbit at 35786 km altitude, matching the rotation period of the Earth. The beam footprint size for GEO ranges from 200 to 3500 km.
- Medium Earth Orbiting (MEO), characterized by a circular orbit varying between 7000 and 25000 km altitude, with a beam footprint size ranging from 100 to 1000 km.
- LEO, which revolves around Earth in a circular orbit between 300 and 1500 km altitude, with a beam footprint size ranging from 100 to 1000 km.

LEO and MEO satellites are also referred to as Non-Geostationary Earth Orbit (NGSO) satellites due to their orbits, which take less time than the Earth's rotation period, typically ranging from 1.5 to 10 h.

The airborne category includes platforms such as Unmanned Aircraft Systems (UAS), typically positioned between 8 and 50 km above ground level, which also encompass HAPs at an altitude of 20 km. Like GEO satellites, UAS also maintains a fixed position relative to a specific point on the ground. The beam footprint size for UAS ranges from 5 to 200 km.

Airborne and spaceborne platforms can fall into two distinct configurations based on their carried payload. NTN platforms utilize either a transparent or regenerative payload setup. In the transparent or bent pipe payload configuration, essential functions like frequency conversion, amplification, and radio frequency filtering are performed on the satellite or UAS platform. Conversely, in the regenerative payload configuration, the NTN platform executes all functions of gNB onboard.

Apart from space and airborne platforms, the NTN access comprises several components:

- NTN terminal refers to either a specialized satellite terminal or 3GPP, UEs. Very small aperture terminals operate within the Ka-band radio frequency (30 GHz in the uplink and 20 GHz in the downlink), while handheld terminals operate within the S-band radio frequency (2 GHz).
- NTN gateway serves as a logical node that connects the NTN platform with the 5G core network.
- Service link denotes the radio connection between the NTN terminal and the NTN platform.
- Feeder link represents the radio connection between the NTN gateway and the NTN platform.

2.3 Integrated TN and NTN Architecture

TN and NTN integration aims to combine the reliability and widespread availability of terrestrial networks with the extensive coverage and unique capabilities of NTNs, such as satellites and high-altitude platforms, to provide seamless connectivity across even the most challenging environments. The Fig. 1 illustrates the architecture of an integrated communication system that connects terrestrial and non-terrestrial networks to provide comprehensive coverage. At the highest level, satellites are shown to be interconnected via "Inter-Satellite Links," which allow for communication directly between satellites, creating a network in space.

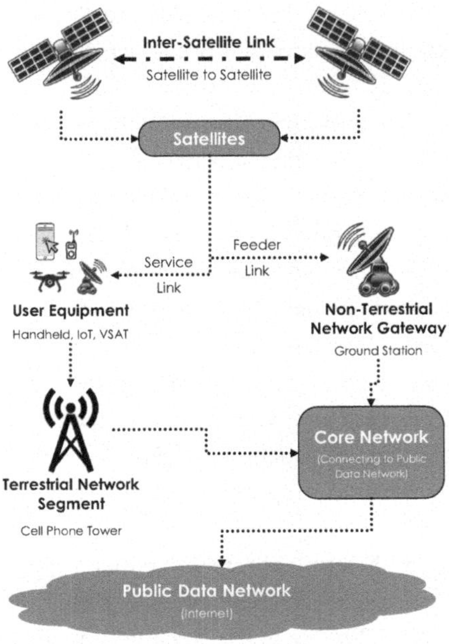

Fig. 1 Integrated communication system architecture merging terrestrial and non-terrestrial networks for global connectivity

These satellites provide two types of links to Earth:

- **Service** Links directly connect satellites to user equipment such as handheld devices, IoT devices, and Very Small Aperture Terminals (VSATs), facilitating direct satellite communication.
- **Feeder Links** connect satellites to a "Non-Terrestrial Network Gateway," which is typically a ground station. This ground station acts as a relay, linking the space-based elements of the network to the terrestrial infrastructure.

On the ground, the "Terrestrial Network Segment," represented by a cell phone tower, provides local coverage. This network segment is connected to the "Core Network," which serves as the central part of the network managing data and calls, interfacing with the broader "Public Data Network" (Internet).

The interconnectedness of satellites, ground stations, terrestrial network segments, and the core network suggests a seamless integration allowing data to be routed through the most appropriate channels, whether they are ground-based or space-based, to ensure optimal connectivity for the end-users.

3 NTN Use Cases

NTNs are poised to play a pivotal role in 5G and beyond systems, catering to diverse sectors such as transport, eHealth, energy, automotive, public safety, and others. The use cases for 5G NTN applications can be classified into three main categories:

- Service Continuity: Ensuring NTN access where terrestrial networks are unfeasible.
- Service Ubiquity: Improving NTN availability during disasters causing temporary terrestrial network outages or damage.
- Service Scalability: Offloading traffic from terrestrial networks, particularly during peak usage hours.

Within the 5G and beyond the framework, NTN supports the International Telecommunication Union (ITU) defined usage scenarios: Massive Machine Type Communications (mMTC), uRLLC, eMBB. While providing uRLLC services might pose challenges due to satellite propagation delays and stringent reliability, availability, and latency requirements, NTN primarily focuses on eMBB and mMTC as the main 5G service facilitators.

For eMBB services, NTN aims to deliver broadband connectivity in underserved areas, on mobile platforms (e.g., vessels, aircraft), and bolster network resilience by combining terrestrial and NTN systems. Additionally, NTN is leveraged to relieve terrestrial networks by providing a broadcast channel for delivering broadcast/multicast content or public safety messages to handheld or vehicle-mounted UEs.

Regarding mMTC, NTN have the ability of connectivity for both local and wide area IoT services. It enables connectivity between the NTN platform and IoT devices for telematics applications (e.g., automotive, energy, agriculture) across wide-area IoT services. For local area IoT services, NTN establishes connectivity between the mobile core network and gNBs serving IoT devices, aggregating information from sensors deployed within one or more cells.

NTN's relevance to 5G New Radio (NR) systems lies in its ability to enhance 5G performance in urban and rural areas, delivering improved data rates and reliability, and providing connectivity in underserved regions for mMTC devices and users.

In specific use-cases, NTN emerges as an advantageous solution for the maritime sector. Given the limitations of terrestrial networks in providing coverage at sea due to cost and capacity constraints, NTN becomes instrumental in facilitating communication within the maritime industry. It manages maritime space, offering seamless sea traffic services to devices and users in collaboration with seaborne platforms. Moreover, NTN can aid in sending notifications and handling emergency requests to enhance maritime safety.

4 NTN in 5G Systems

Up until a few decades ago, satellite and terrestrial networks were perceived as separate entities progressing independently. However, with the advent of current-generation wireless technology, notably 5G, a shift in perspective towards these networks has emerged. The 3GPP standardization has already finalized initial 5G NR specifications and is advancing solutions to support the integration of NTN into 5G NR systems [8]. Initiatives such as SAT5G [9], a part of the H2020 5G Public-Private Partnership (PPP) [10], aim to propose cost-effective solutions for ubiquitous 5G connectivity, thus fostering new prospects within the global 5G market.

A pivotal requirement in this integration involves ensuring service continuity between the 5G NTN Next-Generation Radio Access Network (NG-RAN) and either the 5G NR terrestrial RAN or another 5G NTN NG-RAN [11]. This continuity necessitates specifications that facilitate seamless handover between systems, avoiding service interruptions, and enabling smooth Idle mode UEs operation for optimal network selection.

The collaboration of the NTN segment with terrestrial networks assumes a crucial role in achieving global coverage. This collaboration enhances capacity through precoding techniques and high-frequency reuse, ensuring uninterrupted service even while on the move.

In Ref. [12], discussions revolve around the architectural and technical aspects concerning 5G systems, encompassing NTN. Meanwhile, [13] evaluates the impact of integrating NTN into mobile systems by conducting an experimental comparison based on Key Performance Indicators (KPIs).

The convergence of non-terrestrial and terrestrial networks emerges as an appealing solution in the advancement of 5G technology. Over the recent years, numerous research endeavors have explored the amalgamation of two distinct radio access networks. Notably, [14] pioneers a comprehensive review of Space-Air-Ground Integrated Networks (SAGIN), leveraging deep learning techniques to enhance system performance, specifically in traffic balancing [15].

Moreover, [16] introduces a fresh outlook on integrated systems, delving into Software Defined Space-Terrestrial Integrated Networks, rooted in Software Defined Networking (SDN) principles. This innovative approach segregates the control plane from the data plane, presenting a new dimension to integrated network architectures. In [17] intelligent hybrid network is presented, which is a spectrum efficient self-organizing hybrid terrestrial-satellite backhaul network.

Reference [18] introduces a streamlined approach to integrating non-terrestrial and terrestrial networks by presenting a novel architecture that merges SDN and Network Function Virtualization (NFV) [19]. This implementation involves executing specific hardware functionalities through software mechanisms.

Security stands out as a critical concern in NTN communications. Various studies in the literature have addressed this matter within integrated TN and NTN. Cognitive radio, for instance, has been introduced to enhance the utilization of spectrum when both cellular networks and NTN operate within the same bandwidth. In [20], researchers delved into physical layer security, proposing a stochastic beam-forming approach. Utilizing multi-antenna terrestrial base stations as a means of controlled interference, [21–23] aimed to fortify the security of NTN communications.

Additionally, [24] devised a cooperative secure transmission beam-forming strategy to evaluate communication security in NTN and terrestrial systems. Their focus was on maximizing secrecy rates while adhering to power and transmission quality constraints.

In Ref. [25], an analysis of secrecy performance has been conducted, specifically focusing on connectivity within a multi-antenna NTN directed towards terrestrial recipients (i.e., in the downlink direction). This analysis incorporates multiple cooperative relays while accounting for the presence of various eavesdroppers. Meanwhile, [26] explores diverse adaptive transmission strategies, aiming to mathematically derive the expression for achievable channel capacity within hybrid NTN terrestrial relay networks.

To fortify system security, [27] introduces a joint opportunistic relay selection scheme, enhancing protection against potential attacks. Descriptions of three common attack approaches are outlined in [28], illustrating probable threats to NTN security. Contrary to prior works that employed cooperation in cognitive NTN-terrestrial networks, [29] devises a noncooperative game model with limited information exchange. This model tackles the power control problem in scenarios involving spectrum sharing between the NTN and TN.

Additionally, investigations into cognitive NTN and terrestrial systems have been conducted. [30] examines system performance through outage analysis considering interference temperature constraints, while [31] analytically derives outage probability and ergodic capacity. The exploration of full cooperative relay protocols (such as amplify-forward and decode-forward) is presented in [32], whereas [33] assesses system performance utilizing a partial relay selection scheme.

The 5G wireless technology introduces non-orthogonal multiple access (NOMA) as a key radio access technique, differing from traditional Orthogonal Multiple Access (OMA) methods where each user is served on a distinct orthogonal carrier [34]. Numerous studies in the literature have delved into both NTN and integrated NTN-TN, focusing on NOMA techniques. For instance, [35] presents a survey on multi-satellite cooperative transmission systems, emphasizing NOMA-based multi-satellite relay transmission systems.

In [36], the formulation of achievable ergodic capacity specifically addresses a NOMA-based uplink NTN. Similarly, [37] explores both ergodic capacity and outage probability within a hybrid NTN-terrestrial relay network, employing cooperative NOMA strategies in the downlink direction. Moreover, [38] analyzes and derives the outage probability in a closed form.

Due to interference between terrestrial and NTN systems, especially when the two downlink channels share the same bandwidth, [39] calculates the capacity of the NOMA-base terrestrial-NTN system. Additionally, [40] proposes an optimization design for NTN multicast communications sharing mmWave spectrum with terrestrial communications, leveraging NOMA techniques.

In connection with this topic, both GEO, High Throughput Satellites (HTS) and LEO satellite mega-constellations are anticipated to draw considerable attention from telecommunication operators and researchers. Specifically, GEO and HTSs, with their capability to achieve exceptionally high data rates, facilitate the provision of eMBB services, especially in further enhanced Multimedia Broadcast Multicast Services (FeMBMS) mode [41]. Conversely, LEO satellites support extremely low-latency 5G services, such as uRLLC, owing to the reduced propagation delay in LEO transmissions. Consequently, both GEO and NGSO satellites might be utilized independently or in combination with terrestrial cellular systems in either standalone or non-standalone radio access scenarios.

The work presented in [42] focuses on a dedicated GEO satellite NG-RAN aimed at delivering diverse video services in upcoming 5G NR deployments. This innovative Radio Resource Management (RRM) approach is designed to optimize resource allocation, catering to multiple multimedia video streams. Additionally, [43] introduces a path-oriented network coding technique proposed to enhance reliability and time-effective traffic distribution within NTN and terrestrial mobile systems.

In [44], the concept of a dedicated LEO NG-RAN for 5G mMTC services is explored. The study outlines an uplink scheduling technique aimed at mitigating the impact of differential Doppler shifts on mMTC devices.

However, integrating LEO satellites with 5G technology poses challenges due to distinctive LEO characteristics such as high-speed mobility around Earth, the Doppler effect, and smaller coverage compared to GEO satellites. These challenges necessitate the development of LEO constellations to provide global coverage. [45] introduces a network architecture specifically designed for dense LEO constellations, enhancing reliability and flexibility in integrated NTNs and terrestrial systems.

Within a constellation, LEO satellites are interconnected via Inter-Satellite Links (ISLs). Leveraging the onboard processing capabilities of regenerative payload-based LEOs enables direct data transmissions between these satellites. [46] presents analytical models aimed at determining call blocking and handover failure probabilities within a constellation of regenerative payload-based LEOs. In contrast, transparent payload-based LEOs necessitate routing data traffic to the terrestrial network, resulting in vertical handover scenarios.

Enabling uninterrupted connection transfers across varying wireless access technologies poses a significant challenge for seamless handovers. Positioning-based strategies to minimize latency and manage inter-satellite handovers in satellite communications have been explored in [47]. This involves selecting the nearest satellite during handover events. Meanwhile, [48] presents stochastic and deterministic optimization problems aimed at facilitating handover in diverse aeronautical networks with an SDN controller.

The advent of 5G is characterized not only by the exponential demand for high data-rate services but also by a vast number of interconnected devices. This surge in connected devices, including heterogeneous ones like those within the IoT, has led to the emergence of a new paradigm termed the Internet-of-Space Things (IoST). Introduced in [49], IoST envisions global connectivity, overcoming limitations of terrestrial base stations [50], utilizing low-cost and flexible solutions by amalgamating SDN and NFV paradigms.

NTNs play a pivotal role in supporting broadcast/multicast IoT communications, Internet of Remote Things (IoRT) applications [51], and Internet of Vehicles (IoV) services [52], extending coverage even to remote regions beyond terrestrial reach. Advancements in microelectronics and microsystems industries have propelled the development of smaller yet more powerful satellites, marking a new direction for the forthcoming 5G satellite-era [53]. CubeSats, initially intended for academic and research purposes [54], have evolved over time [55, 56] to become a groundbreaking solution for establishing a global IoST network due to their small payload sizes, cost-effectiveness (design, construction, launch), and scalability [57].

In addition to SDN and NFV [62, 63], 5G introduces concepts like Network Slicing [64] and Edge Computing (EC) [65]. Network Slicing enables scalability, higher availability, and resource optimization by tailoring specific network capabilities based on service requirements. Meanwhile, EC relocates computing and storage resources closer to users, reducing latency. These concepts have been adopted for 5G satellite networks in [66–68].

[66] proposes 5GsatEC, a framework integrating 5G satellite edge computing, optimizing hardware resources and offering a software framework based on microservices for different services and users. [67] introduces edge computing in space-based cloud-fog satellite network slices, reducing delays by incorporating edge computing nodes into satellite network computing architectures. [68] explores CubeSat integration into multi-tenant scenarios by designing an SDN/NFV IoT platform based on EC, including CubeSat constellations.

In summary, 5G technology foresees NTN integration to expand terrestrial coverage and provide advanced services when traditional cellular networks face overload or unavailability. Table 1 classify related literature based on common subjects and open research topics in the 5G and beyond domains.

Table 1 Open research topics in non-terrestrial networks (NTNs)

Topic	Description
Network Architecture for NTNs	Integration of NTNs with terrestrial networks (backhaul, fronthaul), Self-configuration and management of NTN networks [58]
Channel Modeling and Characterization for NTNs	Channel modelling for different NTN technologies (e.g., LEO, MEO, HAPs), Impact of atmospheric effects on signal propagation [59]
Mobility Management forNTNs	Efficient handovers between NTNs and terrestrial networks, Mobility prediction and resource allocation for mobile terminals [60]
Coverage and Capacity Analysis for NTNs	Deployment strategies for NTN constellations to ensure global coverage, Resource allocation schemes to maximize network capacity [58]
Security and Privacy for NTNs	Authentication and authorization mechanisms for NTN networks, Securing user data transmission over NTN [61]
Integration of Terrestrial and Satellite Networks	Convergence of satellite and terrestrial networks, traffic distribution optimization and support for converged networks [61]

5 Conclusion

The chapter is written to provide an extensive review and incorporation of terrestrial and non-terrestrial networks, including deployment scenarios, integration difficulties, and many use cases representing a pivotal advancement in global communication infrastructure. The ground-based and aerial or space-based technologies bridge the gap and innovate through seamless delivery and high-capacity connectivity across the globe. The combination of TN with NTN, including satellites, UAVs, and HAPs revolutionizes the access and way of distributing information.

As we move forward, this chapter is also a vital tool for anyone involved in the business of telecommunications since it explores the architecture of integrated TNs and NTN, the inner workings of NTN, and how it integrates and aligns with 5G communication systems. The realization of a seamless, omnipresent connectivity environment is becoming closer as researchers and industry professionals keep addressing problems and coming up with creative solutions, promising a world where connectivity has no boundaries.

References

1. M. A. Jamshed, A. Kaushik, M. Toka, W. Shin, M. Z. Shakir, S. P. Dash, and D. Dardari, "Synergizing airborne non-terrestrial networks and reconfigurable intelligent surfaces-aided 6g iot," *arXiv preprint arXiv:2310.16371*, 2023.
2. M. A. Jamshed, Q. H. Abbasi, and M. Ur-Rehman, "Feasibility of intelligent reflecting surfaces to combine terrestrial and non-terrestrial networks," *Intelligent Reconfigurable Surfaces (IRS) for Prospective 6G Wireless Networks*, pp. 25–40, 2022.
3. M. A. Jamshed, K. Ali, Q. H. Abbasi, M. A. Imran, and M. Ur-Rehman, "Challenges, applications, and future of wireless sensors in internet of things: A review," *IEEE Sensors Journal*, vol. 22, no. 6, pp. 5482–5494, 2022.
4. M. M. Azari, S. Solanki, S. Chatzinotas, O. Kodheli, H. Sallouha, A. Colpaert, J. F. M. Montoya, S. Pollin, A. Haqiqatnejad, A. Mostaani, *et al.*, "Evolution of non-terrestrial networks from 5g to 6g: A survey," *IEEE communications surveys & tutorials*, 2022.
5. S. Kota and G. Giambene, "6g integrated non-terrestrial networks: Emerging technologies and challenges," in 2021 IEEE International Conference on Communications Workshops (ICC Workshops), pp. 1–6, 2021.
6. M. M. Azari, S. Solanki, S. Chatzinotas, and M. Bennis, "Thz-empowered uavs in 6g: Opportunities, challenges, and trade-offs," *IEEE Communications Magazine*, vol. 60, no. 5, pp. 24–30, 2022.
7. M. A. Jamshed, A. Nauman, M. A. B. Abbasi, and S. W. Kim, "Antenna selection and designing for thz applications: suitability and performance evaluation: a survey," *IEEE Access*, vol. 8, pp. 113246–113261, 2020.
8. L. Chiaraviglio, N. Blefari-Melazzi, W. Liu, J. A. Gutiérrez, J. Van De Beek, R. Birke, L. Chen, F. Idzikowski, D. Kilper, P. Monti, *et al.*, "Bringing 5g into rural and low-income areas: Is it feasible?," *IEEE Communications Standards Magazine*, vol. 1, no. 3, pp. 50–57, 2017.
9. M. Shafi, J. Zhang, H. Tataria, A. F. Molisch, S. Sun, T. S. Rappaport, F. Tufvesson, S. Wu, and K. Kitao, "Microwave vs. millimeter-wave propagation channels: Key differences and impact on 5g cellular systems," *IEEE Communications Magazine*, vol. 56, no. 12, pp. 14–20, 2018.

10. T. S. Rappaport, Y. Xing, G. R. MacCartney, A. F. Molisch, E. Mellios, and J. Zhang, "Overview of millimeter wave communications for fifth-generation (5g) wireless networks-with a focus on propagation models," IEEE Transactions on antennas and propagation, vol. 65, no. 12, pp. 6213–6230, 2017.
11. 3GPP "Study on using satellite access in 5g," 3rd Generation Partnership Project (3GPP), 2018.
12. F. Babich, M. Comisso, A. Cuttin, M. Marchese, and F. Patrone, "Nanosatellite-5g integration in the millimeter wave domain: A full top-down approach," IEEE Transactions on Mobile Computing, vol. 19, no. 2, pp. 390–404, 2019.
13. C. Kourogiorgas, A. Z. Papafragkakis, A. D. Panagopoulos, and V. K. Sakarellos, "Cooperative diversity performance of hybrid satellite and terrestrial millimeter wave backhaul 5g networks," in 2017 International Workshop on Antenna Technology: Small Antennas, Innovative Structures, and Applications (iWAT), pp. 46–49, IEEE, 2017.
14. Q. Wu, J. Xu, Y. Zeng, D. W. K. Ng, N. Al-Dhahir, R. Schober, and A. L. Swindlehurst, "A comprehensive overview on 5g-and-beyond networks with uavs: From communications to sensing and intelligence," IEEE Journal on Selected Areas in Communications, vol. 39, no. 10, pp. 2912–2945, 2021.
15. A. Mudonhi, C. Sacchi, and F. Granelli, "Sdn-based multimedia content delivery in 5g mmwave hybrid satellite-terrestrial networks," in 2018 IEEE 29th Annual International Symposium on Personal, Indoor and Mobile Radio Communications (PIMRC), pp. 1–7, IEEE, 2018.
16. X. Artiga, A. Pérez-Neira, J. Baranda, E. Lagunas, S. Chatzinotas, R. Zetik, P. Gorski, K. Ntougias, D. Perez, and G. Ziaragkas, "Shared access satellite-terrestrial reconfigurable backhaul network enabled by smart antennas at mmwave band," IEEE Network, vol. 32, no. 5, pp. 46–53, 2018.
17. G. Ziaragkas, G. Poziopoulou, J. Núñez-Martínez, J. Baranda, I. Moreno, C. Tsinos, S. Maleki, S. K. Sharma, M. Alodeh, and S. Chatzinotas, "Sansa-hybrid terrestrial–satellite backhaul network: scenarios, use cases, kpis, architecture, network and physical layer techniques," International Journal of Satellite Communications and Networking, vol. 35, no. 5, pp. 379–405, 2017.
18. H. Zhang, C. Jiang, J. Wang, L. Wang, Y. Ren, and L. Hanzo, "Multicast beamforming optimization in cloud-based heterogeneous terrestrial and satellite networks," IEEE Transactions on Vehicular Technology, vol. 69, no. 2, pp. 1766–1776, 2019.
19. B. Di, H. Zhang, L. Song, Y. Li, and G. Y. Li, "Ultra-dense leo: Integrating terrestrial-satellite networks into 5g and beyond for data offloading," IEEE Transactions on Wireless Communications, vol. 18, no. 1, pp. 47–62, 2018.
20. M. Xiao, S. Mumtaz, Y. Huang, L. Dai, Y. Li, M. Matthaiou, G. K. Karagiannidis, E. Björnson, K. Yang, I. Chih-Lin, et al., "Millimeter wave communications for future mobile networks," IEEE Journal on Selected Areas in Communications, vol. 35, no. 9, pp. 1909–1935, 2017.
21. F. Khan and Z. Pi, "mmwave mobile broadband (mmb): Unleashing the 3–300ghz spectrum," in 34th IEEE Sarnoff Symposium, pp. 1–6, IEEE, 2011.
22. W. Li, C. L. Law, V. K. Dubey, and J. Ong, "Ka-band land mobile satellite channel model incorporating weather effects," IEEE Communications letters, vol. 5, no. 5, pp. 194–196, 2001.
23. C. Loo and J. S. Butterworth, "Land mobile satellite channel measurements and modeling," Proceedings of the IEEE, vol. 86, no. 7, pp. 1442–1463, 1998.
24. A. Kelmendi, A. Hrovat, M. Mohorčič, and A. Švigelj, "Alphasat propagation measurements at ka-and q-bands in ljubljana: Three years' statistical analysis," IEEE Antennas and Wireless Propagation Letters, vol. 20, no. 2, pp. 174–178, 2020.
25. J. P. Choi and V. W. Chan, "Optimum power and beam allocation based on traffic demands and channel conditions over satellite downlinks," IEEE Transactions on Wireless Communications, vol. 4, no. 6, pp. 2983–2993, 2005.

26. J. Lei and M. Á. Vázquez-Castro, "Multibeam satellite frequency/time duality study and capacity optimization," Journal of Communications and Networks, vol. 13, no. 5, pp. 472–480, 2011.
27. M. G. Kibria, E. Lagunas, N. Maturo, H. Al-Hraishawi, and S. Chatzinotas, "Carrier aggregation in satellite communications: Impact and performance study," IEEE Open Journal of the Communications Society, vol. 1, pp. 1390–1402, 2020.
28. E. Lagunas, S. Maleki, L. Lei, C. Tsinos, S. Chatzinotas, and B. Ottersten, "Carrier allocation for hybrid satellite-terrestrial backhaul networks," in 2017 IEEE International Conference on Communications Workshops (ICC Workshops), pp. 718–723, IEEE, 2017.
29. E. Lagunas, S. Chatzinotas, and B. Ottersten, "Fair carrier allocation for 5g integrated satellite-terrestrial backhaul networks," in 2018 25th International Conference on Telecommunications (ICT), pp. 617–622, IEEE, 2018.
30. M. Shaat, E. Lagunas, A. I. Perez-Neira, and S. Chatzinotas, "Integrated terrestrial-satellite wireless backhauling: Resource management and benefits for 5g," ieee vehicular technology magazine, vol. 13, no. 3, pp. 39–47, 2018.
31. Y. Kawamoto, T. Kamei, M. Takahashi, N. Kato, A. Miura, and M. Toyoshima, "Flexible resource allocation with inter-beam interference in satellite communication systems with a digital channelizer," IEEE Transactions on Wireless Communications, vol. 19, no. 5, pp. 2934–2945, 2020.
32. X. Liang, J. Jiao, S. Wu, and Q. Zhang, "Outage analysis of multirelay multiuser hybrid satellite-terrestrial millimeter-wave networks," IEEE Wireless Communications Letters, vol. 7, no. 6, pp. 1046–1049, 2018.
33. K. An and T. Liang, "Hybrid satellite-terrestrial relay networks with adaptive transmission," IEEE Transactions on Vehicular Technology, vol. 68, no. 12, pp. 12448–12452, 2019.
34. H. Al-Hraishawi, N. Maturo, E. Lagunas, and S. Chatzinotas, "Scheduling design and performance analysis of carrier aggregation in satellite communication systems," IEEE Transactions on Vehicular Technology, vol. 70, no. 8, pp. 7845–7857, 2021.
35. F. Guidolin, M. Nekovee, L. Badia, and M. Zorzi, "A study on the coexistence of fixed satellite service and cellular networks in a mmwave scenario," in 2015 IEEE International Conference on Communications (ICC), pp. 2444–2449, IEEE, 2015.
36. V. Icolari, D. Tarchi, A. Guidotti, and A. Vanelli-Coralli, "Genetic inspired scheduling algorithm for cognitive satellite systems," in 2016 IEEE International Conference on Communications (ICC), pp. 1–6, IEEE, 2016.
37. E. Lagunas, S. K. Sharma, S. Maleki, S. Chatzinotas, and B. Ottersten, "Resource allocation for cognitive satellite communications with incumbent terrestrial networks," IEEE Transactions on Cognitive Communications and Networking, vol. 1, no. 3, pp. 305–317, 2015.
38. E. Lagunas, S. K. Sharma, S. Maleki, S. Chatzinotas, and B. Ottersten, "Power control for satellite uplink and terrestrial fixed-service co-existence in ka-band," in 2015 IEEE 82nd Vehicular Technology Conference (VTC2015-Fall), pp. 1–5, IEEE, 2015.
39. E. Lagunas, S. Maleki, S. Chatzinotas, M. Soltanalian, A. I. Pérez-Neira, and B. Oftersten, "Power and rate allocation in cognitive satellite uplink networks," in 2016 IEEE International Conference on Communications (ICC), pp. 1–6, IEEE, 2016.
40. S. K. Sharma, S. Chatzinotas, J. Grotz, and B. Ottersten, "3d beamforming for spectral coexistence of satellite and terrestrial networks," in 2015 IEEE 82nd Vehicular Technology Conference (VTC2015-Fall), pp. 1–5, IEEE, 2015.
41. K. An, M. Lin, J. Guyang, T. Liang, J.-b. Wang, and W.-P. Zhu, "Outage performance for the cognitive broadband satellite system and terrestrial cellular network in millimeter wave scenario," in 2017 IEEE International Conference on Communications (ICC), pp. 1–6, IEEE, 2017.
42. Q. Zhang, K. An, X. Yan, and T. Liang, "Coexistence and performance limits for the cognitive broadband satellite system and mmwave cellular network," IEEE Access, vol. 8, pp. 51905–51917, 2020.

43. D. Peng, Y. Li, S. Chatzinotas, and B. Ottersten, "Hybrid analog-digital precoding for mmwave coexisting in 5g-satellite integrated network," in 2020 IEEE 31st Annual International Symposium on Personal, Indoor and Mobile Radio Communications, pp. 1–6, IEEE, 2020.
44. Z. Xiao, P. Xia, and X.-G. Xia, "Enabling uav cellular with millimeter-wave communication: Potentials and approaches," IEEE Communications Magazine, vol. 54, no. 5, pp. 66–73, 2016.
45. L. Zhang, H. Zhao, S. Hou, Z. Zhao, H. Xu, X. Wu, Q. Wu, and R. Zhang, "A survey on 5g millimeter wave communications for uav-assisted wireless networks," IEEE Access, vol. 7, pp. 117460–117504, 2019.
46. Z. Xiao, L. Zhu, and X.-G. Xia, "Uav communications with millimeter-wave beamforming: Potentials, scenarios, and challenges," China Communications, vol. 17, no. 9, pp. 147–166, 2020.
47. Z. Feng, L. Ji, Q. Zhang, and W. Li, "Spectrum management for mmwave enabled uav swarm networks: Challenges and opportunities," IEEE Communications Magazine, vol. 57, no. 1, pp. 146–153, 2018.
48. W. Khawaja, O. Ozdemir, and I. Guvenc, "Uav air-to-ground channel characterization for mmwave systems," in 2017 IEEE 86th Vehicular Technology Conference (VTC-Fall), pp. 1–5, IEEE, 2017.
49. M. T. Dabiri, H. Safi, S. Parsaeefard, and W. Saad, "Analytical channel models for millimeter wave uav networks under hovering fluctuations," IEEE Transactions on Wireless Communications, vol. 19, no. 4, pp. 2868–2883, 2020.
50. M. Polese, L. Bertizzolo, L. Bonati, A. Gosain, and T. Melodia, "An experimental mmwave channel model for uav-to-uav communications," in Proceedings of the 4th ACM Workshop on Millimeter-Wave Networks and Sensing Systems, pp. 1–6, 2020.
51. A. Colpaert, E. Vinogradov, and S. Pollin, "Aerial coverage analysis of cellular systems at lte and mmwave frequencies using 3d city models," Sensors, vol. 18, no. 12, p. 4311, 2018.
52. S. S. Kalamkar, F. Baccelli, F. M. Abinader, A. S. M. Fani, and L. G. U. Garcia, "Beam management in 5g: A stochastic geometry analysis," IEEE Transactions on Wireless Communications, vol. 21, no. 4, pp. 2275–2290, 2021.
53. E. Vinogradov, H. Sallouha, S. De Bast, M. M. Azari, and S. Pollin, "Tutorial on uav: A blue sky view on wireless communication," arXiv preprint arXiv:1901.02306, 2019.
54. A. Colpaert, E. Vinogradov, and S. Pollin, "3d beamforming and handover analysis for uav networks," in 2020 IEEE Globecom Workshops (GC Wkshps, pp. 1–6, IEEE, 2020.
55. H.-L. Song, Y.-c. Ko, J. Cho, and C. Hwang, "Beam tracking algorithm for uav communications using kalman filter," in 2020 International Conference on Information and Communication Technology Convergence (ICTC), pp. 1101–1104, IEEE, 2020.
56. P. Zhou, X. Fang, Y. Fang, R. He, Y. Long, and G. Huang, "Beam management and self-healing for mmwave uav mesh networks," IEEE Transactions on Vehicular Technology, vol. 68, no. 2, pp. 1718–1732, 2018.
57. K. Heimann, J. Tiemann, S. Böcker, and C. Wietfeld, "On the potential of 5g mmwave pencil beam antennas for uav communications: An experimental evaluation," in WSA 2018; 22nd International ITG Workshop on Smart Antennas, pp. 1–6, VDE, 2018.
58. Y. Fumihiro, M. Munehiro, K. Hisayoshi, and A. Junichi, "Multi-layer non-terrestrial network for beyond 5g/6g mobile communications," NTT Technical Review, vol. 21, no. 8, pp. 23–27, 2023.
59. 5G Americas, "5G And Non-Terrestrial Networks." https://www.5gamericas.org/5g-and-non-terrestrial-networks/, 2024. [Online; accessed 26-March-2024].
60. A. Guidotti, A. Vanelli-Coralli, V. Schena, N. Chuberre, M. El Jaafari, J. Puttonen, and S. Cioni, "The path to 5g-advanced and 6g non-terrestrial network systems," in 2022 11th Advanced Satellite Multimedia Systems Conference and the 17th Signal Processing for Space Communications Workshop (ASMS/SPSC), pp. 1–8, IEEE, 2022.

61. F. Rinaldi, H.-L. Maattanen, J. Torsner, S. Pizzi, S. Andreev, A. Iera, Y. Koucheryavy, and G. Araniti, "Non-terrestrial networks in 5g & beyond: A survey," *IEEE access*, vol. 8, pp. 165178–165200, 2020.
62. Z. Zhang, Q. Zhu, and P. Zhang, "Fast beam tracking discontinuous reception for d2d-based uav mmwave communication," *IEEE Access*, vol. 7, pp. 110487–110498, 2019.
63. M. Gapeyenko, V. Petrov, D. Moltchanov, S. Andreev, N. Himayat, and Y. Koucheryavy, "Flexible and reliable uav-assisted backhaul operation in 5g mmwave cellular networks," *IEEE Journal on Selected Areas in Communications*, vol. 36, no. 11, pp. 2486–2496, 2018.
64. L. Wang, Y. L. Che, J. Long, L. Duan, and K. Wu, "Multiple access mmwave design for uav-aided 5g communications," *IEEE Wireless Communications*, vol. 26, no. 1, pp. 64–71, 2019.
65. Z. Khosravi, M. Gerasimenko, S. Andreev, and Y. Koucheryavy, "Performance evaluation of uav-assisted mmwave operation in mobility-enabled urban deployments," in *2018 41st International Conference on Telecommunications and Signal Processing (TSP)*, pp. 1–5, IEEE, 2018.
66. A. Rahmati, Y. Yapici, N. Rupasinghe, I. Guvenc, H. Dai, and A. Bhuyan, "Energy efficiency of rsma and noma in cellular-connected mmwave uav networks," in *2019 IEEE International Conference on Communications Workshops (ICC Workshops)*, pp. 1–6, IEEE, 2019.
67. J. Zhao, F. Gao, L. Kuang, Q. Wu, and W. Jia, "Channel tracking with flight control system for uav mmwave mimo communications," *IEEE Communications Letters*, vol. 22, no. 6, pp. 1224–1227, 2018.
68. Y. Li, H. Wei, L. Li, Y. Han, J. Zhou, and W. Zhou, "An extensible multi-layer architecture model based on leo-mss and performance analysis," in *2019 IEEE 90th Vehicular Technology Conference (VTC2019-Fall)*, pp. 1–6, IEEE, 2019.

IoT and M2M Applications in Satellite Networks

Adeel Iqbal, Atif Shakeel, Adnan Rashid, and Sung Won Kim

Abstract

This chapter discusses how the Internet of Thing (IoT) and Machine-to-Machine (M2M) technologies integrate with satellite communication systems, underlining how these advancements are transforming industries and improving worldwide connectivity. Internet of Thing (IoT) refers to the interconnection of things or physical objects, while M2M enables direct, human-free communication between devices. Integration of all these three technologies may provide numerous benefits, such as network resiliency, global coverage, and scalability. This integration enables real-time data exchange in remote and challenging environments where transformed procedures like asset tracking, disaster management, and environmental monitoring. It also altered industries like agriculture and energy. The

A. Iqbal (✉) · S. W. Kim
School of Computer Science and Engineering, Yeungnam University, Gyeongsan-si, Republic of Korea
e-mail: adeeliqbal@yu.ac.kr

S. W. Kim
e-mail: swon@yu.ac.kr

A. Shakeel
Department of Electrical and Computer Engineering, COMSATS University Islamabad, Islamabad, Pakistan
e-mail: atif_shakeel@comsats.edu.pk

A. Rashid
Department of Electrical and Information Engineering, Politecnico di Bari, Bari, Italy
e-mail: adnan.rashid@poliba.it

© The Author(s), under exclusive license to Springer Nature Switzerland AG 2025
M. A. Jamshed and A. Nauman (eds.), *Integrated Terrestrial and Non-Terrestrial Networks*, Synthesis Lectures on Communications,
https://doi.org/10.1007/978-3-031-63705-6_2

chapter highlights the role that satellite-based IoT and M2M applications play in providing reliable communication, closing the digital divide, and enabling access in underserved areas. Future advancements in the industry, such as Low Earth Orbit (LEO) satellite constellations, AI integration, and quantum communication, have a great deal of promise. Moreover, revolutionized industries like agriculture and energy, and processes such as environmental monitoring, asset tracking, and disaster management. The chapter emphasizes the significance of satellite-based IoT and M2M applications in bridging the digital divide, empowering underserved regions with connectivity, and offering robust communication. Future developments, such as LEO satellite constellations, artificial intelligence integration, and quantum communication, hold immense promise for further enhancing the field. Additionally, energy harvesting, multi-modal connectivity, and collaborative satellite networks offer new dimensions to the realm of connected and intelligent systems. Ultimately, satellite-based IoT and M2M applications are poised to drive socio-economic development, improve safety, and enable sustainable practices across various sectors. With emerging technologies reshaping the future of connectivity and exploration, this chapter outlines the transformative potential of satellite-based IoT and M2M applications in fostering a more connected and intelligent world.

Keywords
Internet of Things • M2M • Satellite communication • Non-terrestrial networks

1 Introduction

Internet-of-Things (IoT) and Machine-to-Machine (M2M) technologies are interrelated concepts that hold huge significance in the modern world. IoT involves a vast network of physical objects embedded with sensors and connectivity capabilities to collect and exchange data over the Internet [1, 2]. These objects, ranging from household appliances to industrial machinery, interact with their environment and other connected devices, enabling specific tasks and valuable insights. The key features of IoT include connectivity, data collection through sensors, data processing, and automation. Various data link layer communication technologies such as Low-Rate Wireless Personal Area Network (Lr-WPAN), Low-Power Wide-Area Network (LP-WAN), and satellite links facilitate data exchange, leading to improved efficiency and convenience. The significance of IoT spans multiple domains such as industries, smart cities, and healthcare. It revolutionizes efficiency, predictive maintenance, and supply chain management in industries. Smart cities benefit from IoT by optimizing traffic flow, managing public services, and monitoring environmental conditions. Healthcare leverages IoT-based medical devices and wearables for real-time health monitoring. In homes, IoT devices provide automation and environmental monitoring is improved through IoT sensors that aid in conservation efforts.

M2M enables direct communication between devices without human intervention, facilitating automation and real-time information sharing. It finds applications in telemetry and industrial automation. M2M significantly contributes to supply chain management, energy and utility monitoring, and transportation and fleet management, enhancing efficiency and logistics in different types of industry [1, 3].

Satellite networks are communication systems that utilize satellites orbiting the Earth to transmit multimedia components (i.e., text, audio, and video) as signals over long distances. Their most significant advantage lies in their unique capability to extend connectivity to remote areas, where traditional terrestrial infrastructure is challenging or economically impractical. With global coverage, they reach even the most isolated locations, including remote regions, islands, and areas far from urban centres. Rapid deployment during disasters further augments their value. Additionally, these networks provide crucial redundancy and resilience for terrestrial networks. During natural disasters or outages, satellite links maintain essential communication services, ensuring continuous connectivity and enabling emergency response capabilities [4].

Satellite networks offer scalability, accommodating a wide range of users and supporting various communication services simultaneously. They can transmit large data volumes, making them suitable for broadband internet services in rural areas, empowering residents and businesses with high-speed internet access for education, healthcare, and e-commerce. Satellite networks are essential for supporting IoT and M2M applications in remote locations. Real-time data exchange and monitoring of assets, environmental conditions, and infrastructure benefit industries like transportation, logistics, and energy, which operate in remote and scattered areas. Furthermore, facilitating telemedicine services and supporting distance learning programs further contribute to improved healthcare and education outcomes in underserved regions. Moreover, satellite phones offer reliable connectivity to individuals in remote areas, including travellers and adventurers [5].

The integration of IoT and M2M technologies with satellite networks is driven by the overarching goal of enhancing global connectivity, particularly in remote and underserved regions and facilitating seamless data exchange in challenging environments as shown in Fig. 1. This strategic integration harnesses the distinct strengths of both IoT or M2M and satellite networks to effectively address specific challenges and unlock novel opportunities across diverse industries. One primary motivation is the achievement of global coverage, a unique attribute of satellite networks that can reach areas where terrestrial networks are either unavailable or impractical to deploy. By merging IoT or M2M capabilities with satellite connectivity, businesses, governments, and individuals gain access to communication services and data exchange capabilities regardless of their geographical location, even in the most remote and isolated regions [6].

Connectivity for remote areas represents another key motivation, as IoT or M2M applications often require communication in distant or geographically challenging locations, such as rural areas, forests, deserts, and offshore facilities. Satellite networks offer a reliable means to connect these remote areas, enabling real-time data collection, monitoring,

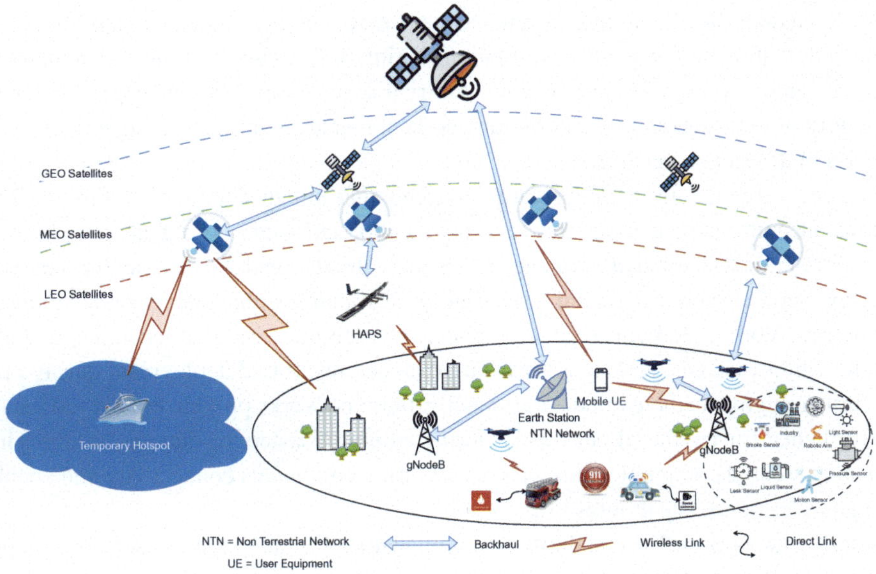

Fig. 1 NTN IoT M2M overview

and control for various applications, including agriculture, environmental monitoring, and resource management. In the context of disaster resilience, satellite networks act as robust backup solutions in the face of natural disasters or emergencies that disrupt terrestrial infrastructure. Moreover, this integration ensures the continuity of critical communication during emergencies, facilitating swift disaster response, coordination, and relief efforts. Seamless connectivity for mobile assets is another driving force, considering IoT or M2M devices are frequently deployed in ships, aircraft, and vehicles. Satellite networks provide continuous connectivity even in transit or remote regions, enabling real-time tracking, monitoring, and data transmission for industries like fleet management, logistics, and transportation [7]. In addition, this integration enhances environmental monitoring capabilities, allowing for the collection of real-time data from remote and inaccessible areas, such as Earth observation, climate monitoring, and wildlife conservation efforts. In the maritime and aviation sectors, where IoT or M2M applications are increasingly deployed to improve safety, navigation, and operational efficiency, the integration with satellite connectivity ensures continuous communication and data exchange over vast oceans and airspace. In remote IoT ecosystems, such as offshore oil rigs, mining sites, or remote research stations, satellite networks provide the essential connectivity for IoT or M2M devices to function efficiently and transmit data to central control centres or cloud platforms. As space exploration advances, integrating IoT or M2M technologies with satellite networks becomes crucial for enabling communication with space probes, rovers, and future human colonies on other planets [8].

2 IoT and M2M Fundamentals

The network of linked items and devices that are embedded with sensors, software, and networking features that enable them to gather, share, and act upon data on their own is known as the Internet of Things, or IoT. IoT makes it possible to combine the digital and physical worlds seamlessly. This makes it possible to monitor, analyze, and operate a variety of systems and processes in real-time across industries, which eventually promotes productivity, efficiency, and creativity.

The IoT system is comprised of several key components that work together to enable connectivity, data collection, processing, and action. Here are the main components of an IoT system:

(1) **Sensors and Actuators**: Sensors gather data from the physical world, such as temperature, humidity, light, motion, and more. Actuators are mechanisms that can act upon the physical world based on data received from sensors, such as turning on/off a light, controlling a valve, or adjusting a thermostat.
(2) **Connectivity**: It enables communication between IoT devices and the Internet or other devices within the IoT ecosystem. This includes different wireless and wired communication technologies such as Lr-WPAN, LP-WAN, cellular networks, and wired communication media.
(3) **IoT Gateways**: They serve as intermediaries between IoT devices and the cloud or central server. They collect data from multiple devices and perform pre-processing, protocol translation, and data filtering before transmitting the data to the cloud for further analysis and storage.
(4) **Cloud Infrastructure**: Cloud-based platforms provide the necessary infrastructure for storing, processing, analyzing, and managing the vast amounts of data generated by IoT devices. Cloud platforms offer scalability, flexibility, and accessibility for IoT applications and services.
(5) **IoT Platforms and Middleware**: They offer tools and services for managing, orchestrating, and integrating IoT devices and data. They provide functionalities such as device management, data analytics, real-time monitoring, security, and application development.
(6) **Data Analytics and Machine Learning**: Algorithms for data analytics and machine learning are used to derive insights, patterns, and trends from the data collected by IoT devices. These insights can be used to optimize processes, make informed decisions, predict outcomes, and automate actions.
(7) **User Interface and Applications**: They enable users to interact with IoT systems, visualize data, set preferences, and receive alerts or notifications. These interfaces can be web-based dashboards, mobile applications, or desktop applications.
(8) **Security Mechanisms**: It is a critical component of IoT systems to protect IoT devices and their generated data privacy, integrity, and confidentiality. It includes encryption,

authentication, access control, secure protocols, firmware updates, and intrusion detection systems.

By integrating these components, IoT systems enable the creation of smart environments, intelligent automation, data-driven decision-making, and innovative services across various industries and domains.

Alternatively, M2M communication involves direct communication between devices or machines without human intervention. It facilitates automated processes and real-time information sharing. Both IoT and M2M aim to enhance efficiency, enable remote monitoring, and support various applications across industries, contributing to the advancement of smart technologies and digital transformation [9]. Like IoT communication system, M2M communication also involves several components working together to enable data exchange between devices. We enlist here the key components involve in M2M communication:

(1) **Sensors and Actuators**: Devices that capture data (sensors) and perform actions (actuators) based on received instructions. They are fundamental for data collection and response in M2M systems.
(2) **Communication Protocols and Technologies**: Standards and technologies governing data exchange between M2M devices. Examples include Message Queuing Telemetry Transport (MQTT), Constrained Application Protocol (CoAP), and various wireless communication protocols.
(3) **Device Connectivity**: Any kind of wired or wireless technologies enable M2M devices to connect to networks to exchange data.
(4) **Data Transmission and Processing**: Processes for sending and receiving data between M2M devices, as well as analyzing and processing data for meaningful insights and actions.
(5) **Security and Authentication**: Measures to protect M2M communication, including encryption, authentication, access control, and secure device management.
(6) **Scalability and Interoperability**: Ability of M2M systems to accommodate growing device numbers and ensure seamless communication between heterogeneous devices and platforms.
(7) **Data Analytics and Visualization**: Techniques and tools for analyzing and visualizing data generated by M2M devices to derive insights and support decision-making.
(8) **Real-Time Monitoring and Control**: Capability to monitor M2M devices and processes in real-time, enabling immediate response and control actions.
(9) **Regulatory Compliance and Standards**: Adherence to regulations, standards, and protocols governing M2M communication to ensure legal compliance and interoperability.
(10) **Applications and Use Cases**: Diverse industries and scenarios where M2M communication is applied, including smart cities, healthcare, transportation, and industrial automation.

While there are similarities between IoT and M2M, they have distinct focuses and characteristics, as shown in Fig. 2. IoT typically involves a more extensive network of interconnected devices and emphasizes the creation of a smart ecosystem where data analytics and automation play a significant role. IoT devices are often designed to be more versatile, and capable of interacting with users and other connected devices to provide personalized experiences and insights. On the other hand, M2M communication tends to focus more on specific use cases and industries, such as industrial automation, telemetry, and remote monitoring. M2M devices may lack the same level of human interaction and user interfaces found in consumer-oriented IoT devices [9, 10]. The key components of both IoT and M2M systems are common, including sensors, actuators, communication protocols, and data processing. Sensors are fundamental devices that detect and measure physical parameters, converting real-world signals into digital data for further processing. Conversely, actuators receive commands or instructions from the central system or other connected devices and carry out physical actions in response. They enable devices to interact with the physical world based on the data received from sensors. Communication protocols encompass a set of guidelines dictating the exchange of data and interactions among devices within an IoT or M2M system. Various communication protocols are employed in these systems, such as MQTT, CoAP, Routing Protocol for Low-Power and Lossy Networks (RPL), IPv6 over Low power Wireless Personal Area Network (6LoWPAN), IPv6 over Low power Wireless Personal Area Network—Neighbor Discovery Protocol (6LoWPAN-NDP), and other link layer technologies based on Lr-WPAN, LP-WAN, cellular networks, and satellite communication. Lastly, data processing involves the analysis, interpretation, and transformation of

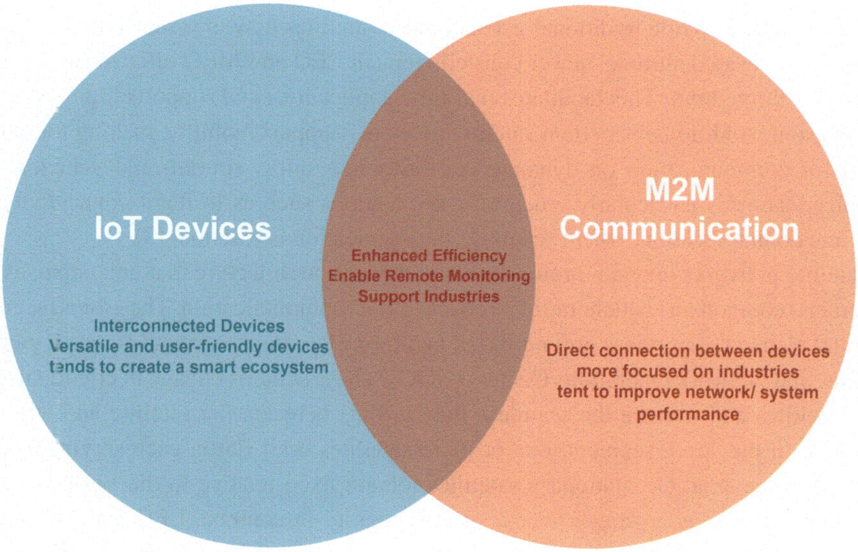

Fig. 2 The similarities and differences between IoT and M2M

raw data collected from sensors. It can occur at the edge within the devices themselves (edge computing) or in the cloud on centralized servers (cloud computing). Data processing helps extract valuable insights, trigger appropriate actions, and optimize system performance [10, 11].

3 Satellite Networks and Their Advantages

Satellite networks possess several distinctive features that contribute to their widespread utility. One notable characteristic is their expansive coverage, extending across vast geographical areas, including remote and challenging terrains. This wide coverage is made possible by satellites in three distinct orbits: geostationary, Medium Earth Orbit (MEO), and Low Earth Orbit (LEO), each designed to fulfil a specific set of communication needs. Global connectivity is another key attribute of satellite networks, facilitating communication between any two points on Earth, regardless of distance. This attribute makes satellite networks invaluable for international communication, emergency response efforts, and providing connectivity to isolated and underserved regions. Reliability is a hallmark of satellite networks, as they demonstrate resilience against physical disruptions like natural disasters or damage to terrestrial infrastructure. The redundancy inherent in satellite links ensures continuous communication, positioning them as robust backups for terrestrial networks. The broadcast capability of satellites is a notable feature, enabling a single transmission to reach multiple receivers simultaneously. This quality proves advantageous for broadcasting television, radio, and other multimedia services to a wide audience [4].

In terms of performance, satellite networks exhibit low latency and high throughput, as shown in Table 1. While traditional geostationary satellites may experience higher latency due to their elevated altitude, newer constellations in LEO and MEO offer reduced latency and higher throughput. This facilitates real-time applications and supports high-speed data transmission. Additionally, certain satellite networks support mobility, making them well-suited for communication on dynamic platforms like ships, aircraft, and vehicles. This mobility feature is particularly crucial for applications such as in-flight connectivity and maritime communication, where seamless communication is essential [3].

Satellite networks operate through a comprehensive architecture that encompasses several crucial components collaborating to enable global communication. The ground segment, a pivotal element, comprises ground-based facilities like satellite ground stations, gateway stations, and Network Operations Centers (NOC). These stations engage in communication with satellites and oversee the seamless flow of data between the satellite and terrestrial networks. In the space segment, one or more satellites orbit Earth, each serving a designated coverage area. Geostationary satellites remain fixed relative to the Earth's surface, providing continuous coverage over a specific region. In contrast, LEO and MEO satellites move about the Earth, necessitating a constellation of satellites to ensure uninterrupted coverage. User terminals, also known as satellite receivers or transceivers, serve as the com-

Table 1 Summary of satellite networks and advantages

Aspect	Description
Coverage	Expansive coverage across different orbits (geostationary, MEO, LEO)
Global connectivity	Communication between any two points on Earth
Reliability	Resilience against physical disruptions and redundancy
Broadcast capability	Simultaneous transmission to multiple receivers
Performance	Low latency and high throughput
Mobility support	Suitable for dynamic platforms like ships, aircraft, and vehicles
Architecture	Ground segment, space segment, user terminals, satellite control system, network management system
Recent developments	Mega-constellations in LEO for global, low-latency internet services
Advantages	
Global coverage	Reaching remote and inaccessible regions
Resilience and redundancy	Backup solution during emergencies and network disruptions
Scalability	Accommodating a large number of users and services
Rapid deployment	Quick implementation in regions lacking communication facilities
Ubiquitous connectivity	Seamless communication across geographical areas
Bridging the digital divide	Providing internet access to remote and rural areas
Versatility	Supporting various communication services and industries
Interconnectivity	Connecting remote assets for real-time data exchange
Satellite constellations	
Geostationary	Fixed position, continuous coverage over specific regions
MEO	Improved coverage and reduced latency
LEO	Significantly reduced latency, ideal for IoT and M2M applications
Hybrid constellations	Combining satellites from different orbits for versatile solutions

munication interface for end-users, whether fixed, like satellite TV dishes on rooftops, or mobile, such as satellite phones or connected devices in vehicles. Various communication protocols are employed by satellite networks to facilitate data exchange between the ground and space segments, outlining the transmission, reception, and processing of data packets. Critical to the system is the satellite control system, responsible for monitoring and controlling satellite operations to maintain their designated orbits and performance levels. Simultaneously, the network management system oversees the entire satellite network, handling resources, optimizing data flow, and ensuring seamless operation and performance. Recent developments in satellite network architecture have witnessed a surge in interest regarding mega-constellations of small satellites in LEO. These constellations, comprising thousands of interconnected satellites, aim to deliver globally accessible low-latency and high-bandwidth internet services. This evolving landscape reflects the continuous adaptation of satellite network architecture, driven by technological advancements and innovative solutions shaping the way we establish global connectivity.

Satellite networks serve as essential components in our interconnected world, offering a range of substantial advantages. Chief among these is their unparalleled ability to provide global coverage, extending communication services to even the most remote and inaccessible regions, including rural areas, islands, and places distant from traditional terrestrial infrastructure. This feature is pivotal for reaching underserved and isolated populations worldwide. A crucial benefit is the resilience and redundancy that satellite networks provide to the overall communication infrastructure. In the face of natural disasters, network outages, or disruptions to terrestrial networks, satellite links act as robust backup solutions, ensuring the uninterrupted provision of essential communication services. This capability proves particularly vital during emergency response efforts when terrestrial systems may encounter challenges [5]. Satellite networks demonstrate excellent scalability, capable of accommodating a large number of users and communication services concurrently. As the demand for connectivity continues to surge, these networks can be expanded and upgraded to meet evolving requirements, addressing the needs of the expanding global population and the increasing number of connected devices in the era of the IoT. Rapid deployment is a notable advantage of satellite networks. Compared to the extensive groundwork required for terrestrial infrastructure, implementing satellite networks in regions lacking established communication facilities is a quicker and more straightforward process. This attribute proves invaluable during emergencies and disaster recovery efforts. Satellite networks facilitate ubiquitous connectivity, allowing users to stay connected across various geographical areas without the need to switch between different networks. This seamless connectivity is particularly advantageous for mobile assets, such as ships, aircraft, and vehicles, which require continuous communication regardless of their location. Moreover, satellites contribute to bridging the digital divide by providing high-speed broadband internet access to remote and rural areas. This empowerment extends to communities gaining access to online education, healthcare services, e-commerce, and a wealth of information resources [12]. The versatility and multi-functionality of satellite networks are evident in their ability to support a wide range of communication services, including voice, data, video, broadcasting, and IoT applications. These networks serve diverse industries, including telecommunications, broadcasting, agriculture, environmental monitoring, and disaster management. Furthermore, satellite networks facilitate the interconnectivity of remote assets such as oil rigs, mining sites, weather stations, and research outposts. This connectivity enables real-time data exchange, monitoring, and control, enhancing efficiency and safety in various sectors [13].

Satellite constellations, comprised of multiple satellites working collaboratively, are structured to offer specific coverage and capabilities, with various orbits, deployment strategies, and characteristics tailoring them for diverse applications, including the IoT and M2M communication [14]. The geostationary satellite constellations position satellites approximately 35,786 km above the Earth's equator, providing continuous coverage over specific regions due to their fixed position. While well-suited for applications requiring constant communication with specific areas, such as broadcasting, television distribution, and fixed satellite communication services, they are less suitable for IoT and M2M applications that demand low latency and high data throughput. The long signal delay caused by the high altitude and limited scalability hinders their effectiveness for widespread IoT and M2M deployments. MEO satellite constellations, operating at altitudes between 2,000 to 20,000 km, offer improved coverage and reduced latency compared to geostationary satellites. While they can support real-time data exchange and provide continuous connectivity to mobile assets, the increased number of satellites required raises complexity and cost. LEO satellite constellations, orbiting at altitudes between 180 and 2,000 km, are particularly well-suited for IoT and M2M applications due to significantly reduced latency, enabling real-time communication. The large number of satellites ensures seamless and continuous global coverage, making them ideal for widespread IoT and M2M deployments, including remote monitoring, asset tracking, environmental sensing, and maritime communication. Hybrid constellations, combining satellites from different orbits, offer versatile solutions for IoT and M2M applications by leveraging the benefits of various orbits. For instance, they may integrate low latency and global coverage from LEO satellites with continuous coverage from geostationary satellites in specific regions. In conclusion, distinct satellite constellations provide varied coverage, latency, and scalability features. While LEO satellite constellations excel in widespread IoT and M2M applications, hybrid constellations and other configurations offer tailored solutions, meeting specific requirements for diverse applications in our connected world [12, 13, 15].

4 IoT and M2M Applications in Satellite Networks

The integration of IoT and M2M communication with satellite networks is revolutionizing various industries by enabling global connectivity, even in remote or inaccessible areas. Here's an overview of how IoT and M2M applications are utilized in satellite networks. Table 2 summarizes this detail.

Table 2 Comparative analysis of IoT and M2M applications in satellite networks

Application	Key points	Common themes/challenges
Agriculture	Precision agriculture with real-time monitoring, IoT sensors optimizing resource utilization, and M2M enhancing automation. Overcomes rural limitations. Targeted interventions reduce costs and environmental impact	Wide coverage in rural areas. Balancing technology with environmental sustainability
Environmental monitoring	IoT-enabled sensors track air quality, monitor ocean temperatures and combat deforestation. Disaster monitoring and early warnings. Real-time data collection and remote environment monitoring	Wide coverage for environmental monitoring. Challenges in balancing technological advancements with sustainability
Asset tracking and logistics	Real-time tracking and optimization of assets, and fleet management benefits from IoT and M2M. Supply chain optimization and enhanced security. Predictive maintenance reduces downtime	Wide coverage for supply chain optimization. Connectivity and security challenges are addressed by real-time tracking
Energy and utilities	Satellite-enabled IoT and M2M are crucial for remote energy infrastructure. Monitoring oil and gas infrastructure, predictive maintenance, IoT in renewable energy	Satellite connectivity is crucial for remote infrastructure. Balancing technological advancements with environmental sustainability
Disaster management	IoT and M2M devices significantly contribute to early warning, efficient coordination, remote monitoring, and recovery efforts	Common benefits of wide coverage and real-time data transmission. Challenges in ensuring reliable satellite communication

4.1 Agriculture

The integration of IoT and M2M technologies with satellite networks has revolutionized the way farmers approach precision agriculture. Precision agriculture involves the use of advanced technologies to optimize crop production and resource utilization, leading to increased efficiency and reduced environmental impact. By leveraging IoT and M2M capabilities through satellite networks, farmers can now monitor their crops, soil conditions, and weather in real-time, enabling data-driven decision-making and proactive management practices. IoT plays a crucial role by incorporating a network of sensors embedded in the fields and on agricultural equipment. These sensors gather real-time data on various parameters, such as soil moisture, temperature, humidity, and nutrient levels. Satellite communication allows seamless transmission of this data to centralized cloud platforms or data centers, where it is processed and analyzed. Farmers can access this valuable information remotely through smartphones or computers, gaining insights into the health and needs of their crops. M2M communication further enhances precision agriculture by facilitating direct communication between devices and machinery in the field. For instance, IoT-enabled sensors can relay data to automated irrigation systems, ensuring that crops receive the right amount of water precisely when needed. M2M systems also enable the integration of agricultural machinery, such as tractors and harvesters, with cloud-based systems. This integration allows for remote monitoring and control of the equipment, optimizing their performance and reducing fuel consumption [16].

The use of satellite networks in precision agriculture overcomes the limitations of traditional terrestrial networks, especially in rural and remote farming areas. Satellite connectivity provides reliable and continuous communication, allowing farmers to access data and control their agricultural operations even in areas with limited terrestrial coverage. With real-time data on crop health, soil conditions, and weather patterns, farmers can implement targeted interventions, such as applying fertilizers or pesticides only where and when needed. This not only maximizes crop yield but also minimizes the use of resources, reducing costs and environmental impact. By closely monitoring weather conditions, farmers can also anticipate potential risks, such as extreme temperatures or heavy rainfall, and take preventive measures to protect their crops [17].

4.2 Environmental Monitoring

Environmental monitoring is a critical aspect of understanding and preserving the health of our planet. The integration of IoT-enabled sensors with satellite networks has significantly advanced the field of environmental monitoring, allowing us to track and analyze various environmental parameters in real-time. This comprehensive approach enables researchers, governments, and organizations to make informed decisions and take proactive measures to address environmental challenges effectively [18]. One of the primary applications of IoT-enabled sensors and satellite networks in environmental monitoring is tracking air quality. Air pollution has a profound impact on public health and the environment. IoT sensors can be deployed in urban areas, industrial zones, and near pollution sources to continuously measure air pollutants such as particulate matter (PM), nitrogen dioxide (NO2), sulfur dioxide (SO2), carbon monoxide (CO), and ozone (O3). These sensors collect real-time data on air quality, which is then transmitted through satellite networks to central databases or cloud platforms. The integration of satellite communication ensures wide coverage, enabling the monitoring of air quality in remote or hard-to-reach areas where terrestrial networks may be limited [19]. Satellite networks also play a crucial role in monitoring ocean temperatures, a vital indicator of climate change and its impact on marine ecosystems. IoT sensors can be placed on buoys or other ocean monitoring platforms to measure sea surface temperatures. These sensors collect data at regular intervals, and the information is transmitted through satellite links to monitoring centers. This continuous monitoring allows scientists and researchers to detect trends and anomalies in ocean temperatures, providing valuable insights into the health of marine environments and the potential impact on weather patterns and sea levels [20].

Deforestation is a significant environmental concern, leading to habitat loss, biodiversity decline, and increased carbon dioxide levels in the atmosphere. IoT-enabled sensors combined with satellite networks offer an effective means to monitor and combat deforestation. Remote sensors deployed in forested areas can detect changes in vegetation cover and canopy density, helping identify areas affected by deforestation. Satellite communication facilitates the transmission of this data to authorities and conservation organizations in real-time. By promptly detecting deforestation activities, timely interventions can be implemented to prevent further destruction and promote sustainable forest management practices [21]. Moreover, satellite networks provide valuable support in disaster monitoring and early warning systems. IoT sensors can be deployed in regions prone to natural disasters such as wildfires, landslides, and floods. These sensors can detect critical changes in environmental parameters, such as temperature, humidity, and soil moisture. In the event of an imminent disaster, the data from these sensors can be transmitted via satellite to authorities, enabling early warnings and evacuation plans to mitigate potential risks and save lives [22].

4.3 Asset Tracking and Logistics

In the realm of asset tracking and logistics, the integration of IoT and M2M applications with satellite networks has revolutionized the way businesses manage their operations across vast distances. These technologies enable real-time monitoring, tracking, and optimization of assets, vehicles, and shipments, leading to increased efficiency, reduced costs, and enhanced supply chain visibility [23]. One of the key benefits of integrating IoT and M2M with satellite networks is the ability to track assets in real-time, irrespective of their location. IoT-enabled tracking devices, equipped with GPS and other sensors, are attached to valuable assets, containers, or vehicles. These devices continuously transmit location data and other relevant information through satellite communication to centralized systems. As a result, businesses can monitor the movement and status of their assets in real-time, ensuring visibility and control throughout the entire supply chain [24]. Fleet management is another area significantly improved by IoT and M2M applications in satellite networks. For transportation and logistics companies operating a fleet of vehicles, real-time tracking is crucial for efficient route planning, on-time deliveries, and fuel optimization. IoT-enabled sensors installed in vehicles provide data on fuel consumption, vehicle health, and driver behaviour. Satellite communication ensures that this data is transmitted reliably, even in remote or cross-border areas with limited terrestrial coverage. Fleet managers can access this data through centralized platforms, allowing them to make data-driven decisions to optimize routes, reduce fuel costs, and enhance overall fleet performance [25].

Supply chain optimization is a key focus in logistics, and the integration of IoT and M2M with satellite networks plays a vital role in achieving this goal. By deploying IoT sensors along the supply chain, businesses can monitor and collect data on various parameters, such as temperature, humidity, and vibration, in real-time. This data ensures that sensitive and perishable goods are transported and stored under the right conditions, reducing the risk of spoilage and ensuring product quality. Satellite communication facilitates the transmission of this data to all stakeholders involved in the supply chain, enabling collaboration and timely decision-making [26]. The combination of IoT and M2M with satellite networks enhances security and reduces the risk of theft and loss of assets during transit. Real-time tracking and monitoring allow businesses to identify any deviations from planned routes or unexpected stops, triggering alerts in case of potential theft or tampering. With this level of security and visibility, businesses can respond promptly to any security threats and recover stolen assets more effectively [24]. Furthermore, IoT and M2M applications in satellite networks enable predictive maintenance for assets and vehicles. IoT sensors can monitor the health of equipment and vehicles in real-time, detecting anomalies or potential issues before they escalate into costly breakdowns. By predicting maintenance needs, businesses can schedule maintenance tasks proactively, minimizing downtime, and optimizing the lifespan of their assets [27].

4.4 Energy and Utilities

Satellite-enabled IoT and M2M solutions play a crucial role in monitoring and managing remote energy infrastructure, offering significant benefits for industries such as oil and gas and renewable energy. These technologies provide real-time data collection, remote monitoring, and efficient communication over vast and often challenging geographic areas, enhancing the safety, reliability, and efficiency of energy operations [28]. In the oil and gas industry, remote pipelines and facilities are spread across extensive and remote regions, making traditional terrestrial communication infrastructure inadequate. Satellite-enabled IoT and M2M solutions bridge this connectivity gap, enabling continuous monitoring of critical infrastructure and assets. IoT sensors deployed along pipelines can monitor parameters such as pressure, flow rates, temperature, and leak detection. The sensors collect data at regular intervals and transmit it via satellite links to centralized control centers. This real-time data allows operators to detect potential issues promptly, such as leaks or pressure fluctuations, enabling them to initiate immediate responses to prevent accidents and mitigate environmental impacts [29]. Satellite-enabled IoT and M2M solutions also support predictive maintenance for oil and gas infrastructure. By continuously monitoring the health of equipment and machinery, operators can detect early signs of equipment deterioration or malfunction. This proactive approach allows them to schedule maintenance and replacement activities in advance, reducing downtime and avoiding costly unplanned shutdowns [30].

For renewable energy installations, such as solar and wind farms located in remote areas, satellite-enabled IoT and M2M solutions are essential for efficient monitoring and optimization. IoT sensors placed on solar panels or wind turbines can collect data on energy generation, equipment performance, and environmental conditions. Satellite communication facilitates the transmission of this data to monitoring centers, allowing operators to analyze energy production and identify potential efficiency improvements. This data-driven approach ensures optimal energy output and assists in identifying any maintenance needs in renewable energy installations located far from urban centers [31]. Moreover, in areas with limited terrestrial connectivity, satellite-enabled IoT and M2M solutions enable seamless communication between renewable energy installations and utility management systems. This connectivity allows operators to remotely control energy generation and distribution, ensuring a stable power supply to consumers [32]. Satellite-enabled solutions also contribute to the safety and security of remote energy infrastructure. In addition to monitoring for potential leaks or equipment malfunctions, satellite communication ensures constant communication with personnel working in isolated locations. This connectivity allows for immediate responses to emergencies, enabling timely evacuation or assistance when required [33].

4.5 Disaster Management

The use of IoT and M2M devices in satellite networks has significantly advanced disaster management, enabling early warning systems, efficient disaster response coordination, and effective recovery efforts. These technologies play a critical role in enhancing disaster preparedness and mitigating the impact of natural disasters and emergencies [34]. IoT sensors and M2M devices are deployed in disaster-prone areas to create early warning systems that continuously monitor environmental parameters. These sensors can detect changes in factors such as seismic activity, weather conditions, water levels, and air quality. The data collected by these sensors is transmitted via satellite links to central monitoring centers and disaster management authorities. Real-time data analysis allows for the early detection of potential disasters, such as earthquakes, hurricanes, floods, and tsunamis. Early warning systems can trigger automated alerts and notifications to the public, providing them with crucial information to take necessary precautions and evacuate to safer locations, reducing the risk of casualties and property damage [35]. During a disaster, efficient coordination among various response agencies is essential for effective relief efforts. Satellite-enabled IoT and M2M devices play a vital role in enabling seamless communication and data exchange among emergency responders. IoT sensors can be deployed in disaster-affected areas to monitor the status of critical infrastructure, assess the extent of damage, and track the movement of relief supplies. M2M communication allows for real-time sharing of information and situational updates among different response teams, enabling better coordination and decision-making. Satellite networks ensure that this communication remains operational even in areas where terrestrial communication infrastructure has been disrupted [36].

Remote Monitoring and Assessments: In the aftermath of a disaster, access to affected areas may be challenging due to damage or safety concerns. Satellite-enabled IoT and M2M devices provide a means to remotely monitor and assess the situation in disaster-hit regions. For example, drones equipped with IoT sensors can conduct aerial surveys to assess the extent of damage and identify areas that require immediate assistance. This data is transmitted through satellite links, allowing responders to analyze the situation and prioritize their actions accordingly. The ability to remotely monitor disaster-affected areas helps speed up the response and recovery process and ensures the safety of response teams [34]. IoT and M2M technologies continue to be valuable assets during the recovery and reconstruction phase. Sensors can be deployed to monitor the progress of rebuilding infrastructure and the restoration of essential services. For example, IoT-enabled environmental sensors can monitor air and water quality to ensure that the recovery efforts do not cause further harm to the environment. Satellite networks facilitate the transmission of this data to central control centers, enabling authorities to monitor the recovery process in real-time and make necessary adjustments as required [37].

5 Technical Challenges and Solutions

Integrating IoT and M2M with satellite networks presents several technical challenges that need to be addressed to ensure reliable and efficient communication in remote and challenging environments. These challenges arise due to the unique characteristics of satellite communication, such as long transmission distances, limited bandwidth, and potential latency issues. However, advancements in technology have led to innovative solutions that overcome these challenges and pave the way for seamless integration.

5.1 Power Constraints

IoT and M2M devices deployed in remote and hard-to-reach areas often rely on battery power or energy harvesting sources, making power efficiency a critical concern. Traditional satellite communication systems may consume significant power during data transmission and reception, leading to rapid battery drain in resource-constrained devices. To tackle power constraints, significant research has been devoted to the development of low-power IoT devices. This includes the use of energy-efficient hardware components, optimized communication protocols, and intelligent power management techniques. Energy harvesting technologies, such as solar panels, kinetic energy harvesters, or even radio frequency (RF) energy scavenging, offer sustainable power sources for IoT devices. Additionally, advancements in low-power satellite communication protocols, such as Low Power Wide Area Networks (LPWANs) over satellite links, help minimize power consumption without compromising data integrity [38].

5.2 Bandwidth Limitations

Satellite communication typically has limited bandwidth compared to terrestrial networks, making efficient data transmission crucial for IoT and M2M applications. Transmitting large volumes of data from numerous IoT devices can quickly saturate the available bandwidth, leading to data congestion and delays. Adaptive data compression techniques are employed to optimize data transmission over satellite links. These techniques compress data at the source before transmission and decompress it at the receiving end, reducing the overall data size and conserving valuable bandwidth. Additionally, satellite-based IoT and M2M systems can employ data aggregation strategies, where data from multiple IoT devices is combined before transmission, further reducing the total data payload. Prioritizing critical data and employing Quality of Service (QoS) mechanisms ensure that essential information is given higher priority, improving the overall efficiency of data transfer [39].

5.3 Latency Issues

Satellite communication suffers from latency due to the longer distance data has to travel between Earth and the satellite. High latency can be problematic for real-time applications that require immediate response and feedback. To mitigate latency issues, IoT applications may require the use of LEO satellite constellations. LEO constellations have lower altitudes and, consequently, shorter round-trip signal travel times, reducing latency. Additionally, advancements in satellite payload and onboard processing allow for intelligent edge computing at the satellite ground stations or directly on the satellites. This means that data can be processed closer to the source, reducing the need for data to travel to Earth before being processed, thereby decreasing latency for critical applications [40].

5.4 Coverage and Connectivity

Ensuring seamless and continuous connectivity, particularly in challenging terrains and harsh environmental conditions, is essential for IoT and M2M applications relying on satellite networks. The design of satellite constellations can be optimized to achieve global coverage and provide redundancy. A mix of geostationary satellites for continuous coverage in specific regions and LEO/MEO satellites for broader coverage can be employed. Additionally, the use of inter-satellite links within satellite constellations enables data relay between satellites, improving overall coverage and reducing potential coverage gaps. Moreover, advances in beamforming and antenna technologies allow for improved satellite signal strength and enhanced connectivity in challenging environments [41].

5.5 Network Interoperability

The integration of IoT and M2M devices with satellite networks requires compatibility and seamless communication between different devices and satellite systems. Standardization of communication protocols is vital to ensure interoperability among various IoT devices and satellite networks. The adoption of industry-standard protocols and open-source solutions promotes compatibility and simplifies the integration process. For instance, the use of MQTT (Message Queuing Telemetry Transport) or CoAP (Constrained Application Protocol) as lightweight and efficient communication protocols enables smooth data exchange between IoT devices and satellite gateways[41].

The advancements in technology and innovative solutions have addressed the technical challenges of integrating IoT and M2M with satellite networks. Low-power IoT devices, adaptive data compression techniques, intelligent edge computing, optimized satellite constellation designs, and standardization of communication protocols have contributed to more efficient and reliable satellite-based IoT and M2M applications. As these technologies continue to evolve, they will further enhance global connectivity, provide valuable data insights, and improve decision-making processes in remote and challenging environments. The continuous development and deployment of these solutions contribute to a more connected and intelligent world, empowering industries and communities across the globe [38–41].

6 Security and Privacy Considerations

The importance of security in IoT and M2M applications cannot be overstated, especially when connected devices transmit sensitive data via satellite links. Satellite-based communications introduce unique security challenges due to the long transmission distances and the potential for interception by malicious actors. Safeguarding data in satellite-based communications requires robust encryption methods, authentication mechanisms, and access control to ensure the confidentiality, integrity, and availability of sensitive information [42].

6.1 Encryption Methods

Encryption is a fundamental security measure to protect data transmitted over satellite links. Data encryption ensures that even if intercepted, the data remains unintelligible to unauthorized parties. Advanced encryption algorithms, such as Advanced Encryption Standard (AES) or Rivest-Shamir-Adleman (RSA), are used to secure data at the source before transmission. To safeguard data in satellite-based communications, end-to-end encryption is essential. IoT and M2M devices should encrypt data at the source using strong encryption algorithms. Additionally, secure key management is crucial to ensure that encryption keys are properly generated, stored, and distributed. The use of Hardware Security Modules (HSMs) or Trusted Platform Modules (TPMs) enhances the security of encryption keys, preventing unauthorized access [42].

6.2 Authentication Mechanisms

Authentication ensures that only authorized devices can access the satellite network and transmit data. Unauthorized access could lead to data breaches or tampering with sensitive information. To enforce strong authentication in satellite-based communications, mutual authentication is recommended. Both the IoT or M2M devices and the satellite gateway or ground station should authenticate each other before data exchange. This prevents Man-in-the-Middle (MitM) attacks and unauthorized devices from gaining access. Technologies like digital certificates or cryptographic tokens can be used for secure device authentication [43].

6.3 Access Control Mechanisms

Access control ensures that only authorized users or devices have access to specific data or functionalities within the satellite network. Implementing access control mechanisms in satellite-based communications is essential. Role Based Access Control (RBAC) allows system administrators to define access levels and permissions based on user roles. Access Control Lists (ACLs) can be employed to restrict access to specific resources or data based on predefined rules. Proper configuration of access control mechanisms prevents unauthorized access and unauthorized modifications to data [44].

6.4 Secure Satellite Ground Stations

Securing satellite ground stations is crucial as they serve as gateways between the satellite network and the terrestrial infrastructure. Vulnerabilities at ground stations could compromise the entire satellite communication system. Implementing strong physical and logical security measures at satellite ground stations is vital. Access to ground stations should be restricted and monitored. Encryption and secure communication protocols should be used for data exchange between ground stations and the satellite network. Regular security audits and vulnerability assessments help identify and address potential security risks [1].

6.5 Over-The-Air (OTA) Updates

IoT and M2M devices in remote locations may require firmware or software updates to patch security vulnerabilities or add new features. However, Over-The-Air (OTA) updates can be vulnerable to attacks if not secured properly. Secure OTA update mechanisms are critical to prevent unauthorized and malicious updates. Digitally signed updates and secure boot mechanisms verify the authenticity and integrity of the updates before installation. Ensuring that OTA updates are encrypted during transmission adds a layer of security [45].

Security is paramount in IoT and M2M applications, especially when sensitive data is transmitted via satellite links. Robust encryption methods, mutual authentication, access control mechanisms, and secure satellite ground stations are essential to safeguard data and maintain the integrity of satellite-based communications. As satellite networks continue to play a significant role in global connectivity and the proliferation of IoT and M2M applications, ensuring a secure communication environment is vital to protect against potential threats and cyberattacks [1, 42–45].

7 Case Studies

Real-world examples of successful IoT and M2M implementations in satellite networks showcase the practical benefits and outcomes achieved across various sectors. These case studies highlight the effectiveness of integrating IoT and M2M technologies with satellite networks to solve specific challenges and improve operational efficiency.

7.1 Agriculture: Crop Monitoring and Precision Farming, CropX Improving Water Efficiency in Agriculture

CropX,[1] an agricultural technology company, deployed IoT-enabled soil sensors integrated with satellite communication to monitor soil conditions and optimize irrigation practices. The soil sensors collect data on moisture levels, temperature, and nutrient content in real-time. This information is transmitted via satellite links to a central cloud-based platform. Satellite connectivity ensures seamless data transmission from remote fields, allowing farm-

[1] https://cropx.com/.

ers to make data-driven decisions on irrigation scheduling and precise water application. By optimizing water usage, CropX helped farmers reduce water consumption, increase crop yields, and improve overall resource efficiency [46].

7.2 Environmental Monitoring: Tracking Deforestation and Forest Health, Rainforest Connection Combating Illegal Logging

Rainforest Connection (RFCx)[2] is an organization that uses IoT devices equipped with acoustic sensors to monitor and detect illegal logging activities in rainforests. These devices are deployed in remote forest areas and connected to satellite networks, allowing real-time audio monitoring. When the sensors detect the sound of chainsaws or other illegal logging activities, alerts are sent via satellite links to local authorities, enabling timely response and intervention. The implementation of IoT and M2M technologies via satellite networks has been instrumental in protecting biodiversity, preserving natural habitats, and deterring illegal deforestation [47].

7.3 Asset Tracking and Logistics: Fleet Management and Cargo Monitoring, Fleet Complete Enhancing Fleet Efficiency and Safety

Fleet Complete,[3] a fleet management solutions provider, utilizes satellite-based IoT devices to track and monitor vehicles in remote regions. These devices are equipped with GPS and satellite communication capabilities, enabling real-time tracking and monitoring of fleet operations even in areas with limited cellular coverage. The integration of IoT and M2M technologies with satellite networks has helped fleet operators optimize route planning, monitor driver behaviour, and ensure timely delivery of goods. Additionally, satellite-based communication has improved driver safety by providing communication options in emergencies where terrestrial networks may not be available [48].

7.4 Energy and Utilities: Monitoring Remote Energy Infrastructure, BP Pipeline Integrity Monitoring

BP,[4] a multinational oil and gas company, leveraged IoT and M2M technologies integrated with satellite communication to monitor the integrity of its oil and gas pipelines in remote and challenging environments. IoT sensors were deployed along the pipelines to collect data on pressure, temperature, and other parameters. The collected data is transmitted via

[2] https://rfcx.org/.

[3] https://www.fleetcomplete.com/.

[4] https://www.bp.com/.

satellite links to a central monitoring station, enabling real-time analysis of pipeline health. The implementation of IoT and M2M in satellite networks has helped BP enhance pipeline safety, detect potential leaks or anomalies, and minimize operational downtime [49].

7.5 Disaster Management: Early Warning Systems and Response Coordination, Red Cross Disaster Response and Relief

The Red Cross,[5] in collaboration with various partners, uses IoT devices and M2M communication integrated with satellite links to establish early warning systems for disaster-prone regions. These systems include IoT sensors for monitoring weather conditions, water levels, and seismic activities. Data collected from these sensors is transmitted via satellite networks to central command centers, allowing timely detection and prediction of disasters. The implementation of IoT and M2M in satellite networks has enabled the Red Cross to enhance disaster response coordination, provide timely warnings to at-risk communities, and facilitate efficient relief efforts during emergencies [50].

These real-world case studies demonstrate the successful integration of IoT and M2M technologies with satellite networks, providing practical benefits and positive outcomes across various sectors. From precision agriculture to environmental monitoring, asset tracking, and disaster management, the combination of IoT and M2M in satellite networks empowers industries and organizations to make data-driven decisions, improve operational efficiency, and contribute to sustainability and safety in a connected world.

8 Future Trends and Opportunities

The future of satellite-based IoT and M2M applications holds great promise, driven by advancements in technology and the increasing demand for global connectivity. Several potential developments are on the horizon that could further enhance the integration of these technologies:

8.1 LEO Satellite Constellations

LEO satellite constellations, such as Starlink,[6] OneWeb,[7] and Kuiper,[8] are being deployed to offer low-latency, high-throughput internet services. These constellations consist of thousands of small satellites working together to provide continuous global coverage. In the

[5] https://www.redcross.org/.

[6] https://www.starlink.com/.

[7] https://oneweb.net/.

[8] https://www.aboutamazon.com/what-we-do/devices-services/project-kuiper.

future, as these mega-constellations become fully operational, they will significantly improve the connectivity and data transfer capabilities for satellite-based IoT and M2M applications [4].

8.2 High-Performance Communication Protocols

With the increasing data demands of IoT and M2M applications, there is a need for high-performance communication protocols optimized for satellite links. Emerging protocols, such as Delay-Tolerant Networking (DTN) and Satellite Data Relay Protocol (SDRP), aim to enhance data transmission efficiency, reduce latency, and ensure reliable communication in challenging satellite environments [51].

8.3 Edge Computing in Satellite Terminals

Edge computing, where data processing and analytics occur closer to the data source, will play a significant role in satellite-based IoT and M2M applications. By incorporating edge computing capabilities into satellite terminals or user devices, real-time data processing and decision-making can occur at the edge of the network, reducing the need for extensive data transmission to centralized servers.

8.4 Artificial Intelligence (AI) and Machine Learning (ML) Integration

The integration of Artificial Intelligence (AI) and Machine Learning (ML) into satellite-based IoT and M2M applications will enable more intelligent data analysis and predictive capabilities. AI algorithms can process large volumes of data collected from IoT devices and satellites, providing valuable insights, optimizing operations, and enabling autonomous decision-making.

8.5 Quantum Communication

Quantum communication offers enhanced security and virtually unbreakable encryption. Integrating quantum communication technology with satellite networks can provide Quantum Key Distribution (QKD) for secure data transmission, addressing the increasing security concerns in satellite-based IoT and M2M applications.

8.6 Interplanetary IoT

As space exploration advances, interplanetary IoT is envisioned, where IoT and M2M technologies enable communication with space probes, rovers, and future human colonies on other planets. Integrating satellite-based communication with interplanetary missions will revolutionize space exploration and scientific research.

8.7 Energy Harvesting for IoT Devices

To address power constraints in remote IoT deployments, energy harvesting techniques, such as solar, kinetic, or thermal energy harvesting, can be integrated into IoT devices. This will enable the autonomous and sustainable operation of IoT sensors in remote locations without the need for frequent battery replacements.

8.8 Multi-Modal Connectivity

The future of satellite-based IoT and M2M applications may involve multi-modal connectivity, where devices seamlessly switch between satellite links, terrestrial networks, and other communication technologies (e.g., 5G/6G, LPWAN) based on availability and data requirements. This dynamic connectivity approach will ensure continuous and optimal communication in diverse environments.

8.9 Collaborative Satellite Networks

Future developments may see collaborative efforts among satellite network providers and IoT platform developers to offer tailored solutions for specific industries or use cases. Collaborative satellite networks can pool resources, share data, and optimize satellite coverage, enabling more cost-effective and efficient solutions.

9 Conclusion

The chapter underscores the pivotal role played by satellite-based IoT and M2M applications in the transformation of industries and the enhancement of global connectivity. It delves into the comprehensive overview of IoT and M2M, delineating their functions in connecting physical objects with sensors and facilitating direct communication between devices. These technologies drive digital transformation, revolutionizing industries and reshaping interactions in the global landscape. The discussion emphasizes the benefits of satellite networks, offering global coverage, essential redundancy, and scalability to address diverse connectivity needs. The integration of IoT and M2M with satellite networks is motivated by the goal of extending global connectivity, bridging the digital divide, and facilitating real-time data exchange, particularly in challenging environments and remote areas. Future developments in satellite-based IoT and M2M applications are explored, including emerging technologies such as LEO satellite constellations, advanced communication protocols, edge computing, artificial intelligence/machine learning integration, quantum communication, and interplanetary IoT. Additionally, energy harvesting, multi-modal connectivity, and collaborative satellite networks are anticipated to shape the field's evolution.

The transformative impact of satellite-based IoT and M2M applications is evident across various sectors. In agriculture, precision farming optimizes resource usage, while environmental monitoring tracks crucial parameters for better resource management. Asset tracking, logistics, energy and utilities monitoring, and disaster management all benefit from the integration of these technologies. The significance of satellite-based applications lies in their potential to improve efficiency, increase automation, optimize resource utilization, and enhance decision-making. By extending connectivity to remote areas and enabling real-time data exchange, these applications foster socio-economic development, improve safety, and contribute to sustainability. The chapter concludes by highlighting the promising future of satellite-based IoT and M2M applications, presenting opportunities for advancements in connectivity, security, and interplanetary exploration. As technology evolves, these applications are expected to continue empowering industries, improving global connectivity, and contributing to a more connected and intelligent world.

References

1. M. Centenaro, C. E. Costa, F. Granelli, C. Sacchi, and L. Vangelista, "A survey on technologies, standards and open challenges in satellite iot," *IEEE Communications Surveys & Tutorials*, vol. 23, no. 3, pp. 1693–1720, 2021.
2. M. A. Jamshed, K. Ali, Q. H. Abbasi, M. A. Imran, and M. Ur-Rehman, "Challenges, applications, and future of wireless sensors in internet of things: A review," *IEEE Sensors Journal*, vol. 22, no. 6, pp. 5482–5494, 2022.
3. M. Marchese, A. Moheddine, and F. Patrone, "Iot and uav integration in 5g hybrid terrestrial-satellite networks," *Sensors*, vol. 19, no. 17, p. 3704, 2019.
4. A. Sattarzadeh, Y. Liu, A. Mohamed, R. Song, P. Xiao, Z. Song, H. Zhang, R. Tafazolli, and C. Niu, "Satellite-based non-terrestrial networks in 5g: Insights and challenges," *IEEE Access*, vol. 10, pp. 11274–11283, 2022.
5. B. Al Homssi, A. Al-Hourani, K. Wang, P. Conder, S. Kandeepan, J. Choi, B. Allen, and B. Moores, "Next generation mega satellite networks for access equality: Opportunities, challenges, and performance," *IEEE Communications Magazine*, vol. 60, no. 4, pp. 18–24, 2022.
6. M. Giordani and M. Zorzi, "Non-terrestrial networks in the 6g era: Challenges and opportunities," *IEEE Network*, vol. 35, no. 2, pp. 244–251, 2021.
7. M. M. Azari, S. Solanki, S. Chatzinotas, O. Kodheli, H. Sallouha, A. Colpaert, J. F. Mendoza Montoya, S. Pollin, A. Haqiqatnejad, A. Mostaani, E. Lagunas, and B. Ottersten, "Evo-

lution of non-terrestrial networks from 5g to 6g: A survey," *IEEE Communications Surveys & Tutorials*, vol. 24, no. 4, pp. 2633–2672, 2022.
8. F. Rinaldi, H.-L. Maattanen, J. Torsner, S. Pizzi, S. Andreev, A. Iera, Y. Koucheryavy, and G. Araniti, "Non-terrestrial networks in 5g & beyond: A survey," *IEEE Access*, vol. 8, pp. 165178–165200, 2020.
9. F. Theoleyr and A.-C. Pang, *Internet of things and M2M communications*. River Publishers, 2013.
10. Y. A. Qadri, A. Nauman, Y. B. Zikria, A. V. Vasilakos, and S. W. Kim, "The future of healthcare internet of things: A survey of emerging technologies," *IEEE Communications Surveys & Tutorials*, vol. 22, no. 2, pp. 1121–1167, 2020.
11. A. Rashid, T. Pecorella, and F. Chiti, "Toward resilient wireless sensor networks: A virtualized perspective," *Sensors*, vol. 20, no. 14, 2020.
12. R. Giuliano and E. Innocenti, "Machine learning techniques for non-terrestrial networks," *Electronics*, vol. 12, no. 3, 2023.
13. B. R. Elbert, *Introduction to satellite communication*. Artech house, 2008.
14. G. Geraci, D. López-Pérez, M. Benzaghta, and S. Chatzinotas, "Integrating terrestrial and non-terrestrial networks: 3d opportunities and challenges," *IEEE Communications Magazine*, vol. 61, no. 4, pp. 42–48, 2023.
15. O. Kocheli, E. Lagunas, N. Maturo, S. K. Sharma, B. Shankar, J. F. M. Montoya, J. C. M. Duncan, D. Spano, S. Chatzinotas, S. Kisseleff, *et al.*, "Satellite communications in the new space era: A survey and future challenges," *IEEE Communications Surveys & Tutorials*, vol. 23, no. 1, pp. 70–109, 2020.
16. S. Kota and G. Giambene, "Satellite 5g: Iot use case for rural areas applications," in *Proceedings of the Eleventh International Conference on Advances in Satellite and Space Communications-SPACOMM*, pp. 24–28, 2019.
17. M. Abbasi, M. H. Yaghmaee, and F. Rahnama, "Internet of things in agriculture: A survey," in *2019 3rd International Conference on Internet of Things and Applications (IoT)*, pp. 1–12, IEEE, 2019.
18. S. L. Ullo and G. R. Sinha, "Advances in smart environment monitoring systems using iot and sensors," *Sensors*, vol. 20, no. 11, p. 3113, 2020.
19. T. W. Ayele and R. Mehta, "Air pollution monitoring and prediction using iot," in *2018 second international conference on inventive communication and computational technologies (ICICCT)*, pp. 1741–1745, IEEE, 2018.
20. J. Chen, H. Yan, S. Bao, X. Cui, C. Bai, and H. Wang, "Evaluating the contribution of satellite measurements to the reconstruction of three-dimensional ocean temperature fields in combination with argo profiles," *Acta Oceanologica Sinica*, vol. 41, no. 2, pp. 65–79, 2022.
21. R. Singh, A. Gehlot, S. V. Akram, A. K. Thakur, D. Buddhi, and P. K. Das, "Forest 4.0: Digitalization of forest using the internet of things (iot)," *Journal of King Saud University-Computer and Information Sciences*, vol. 34, no. 8, pp. 5587–5601, 2022.
22. M. Esposito, L. Palma, A. Belli, L. Sabbatini, and P. Pierleoni, "Recent advances in internet of things solutions for early warning systems: A review," *Sensors*, vol. 22, no. 6, p. 2124, 2022.
23. S. Parrino, G. Peruzzi, and A. Pozzebon, "Lopatran: Low power asset tracking by means of narrow band iot (nb-iot) technology," *Sensors*, vol. 21, no. 11, p. 3772, 2021.
24. J. Garg, A. Agrawal, V. Gupta, and S. Dhiman, "Secure asset tracking using gps, mqtt, rfid, gsm and cloud technologies," in *International Conference on Electrical and Electronics Engineering*, pp. 528–536, Springer, 2022.
25. V. S. Padilla, A. Collaguazo, J. J. Cabrera, and J. Pesantez, "M2m technology for bus fleet management. case study: A college transportation system," in *2020 IEEE Technology & Engineering Management Conference (TEMSCON)*, pp. 1–5, IEEE, 2020.

26. L. Davoli, L. Belli, and G. Ferrari, "Iot-enabled smart monitoring and optimization for industry 4.0," *I*nternet of Things: Techonologies, Challenges and Impact; MAT: Roma, Italy, vol. 5, pp. 207–226, 2020.
27. S. Bhanji, H. Shotz, S. Tadanki, Y. Miloudi, and P. Warren, "Advanced enterprise asset management systems: Improve predictive maintenance and asset performance by leveraging industry 4.0 and the internet of things (iot)," in *A*SME/IEEE Joint Rail Conference, vol. 84775, p. V001T12A002, American Society of Mechanical Engineers, 2021.
28. V. Moudgil, K. Hewage, S. A. Hussain, and R. Sadiq, "Integration of iot in building energy infrastructure: A critical review on challenges and solutions," *R*enewable and Sustainable Energy Reviews, vol. 174, p. 113121, 2023.
29. T. R. Wanasinghe, R. G. Gosine, L. A. James, G. K. Mann, O. De Silva, and P. J. Warrian, "The internet of things in the oil and gas industry: a systematic review," *I*EEE Internet of Things Journal, vol. 7, no. 9, pp. 8654–8673, 2020.
30. M. Molęda, B. Młysiak-Mrozek, W. Ding, V. Sunderam, and D. Mrozek, "From corrective to predictive maintenance - a review of maintenance approaches for the power industry," *S*ensors, vol. 23, no. 13, 2023.
31. P. Bedi, S. Goyal, A. S. Rajawat, R. N. Shaw, and A. Ghosh, "Application of ai/iot for smart renewable energy management in smart cities," *A*I and IoT for Smart City Applications, pp. 115–138, 2022.
32. G. B. Gaggero, M. Marchese, A. Moheddine, and F. Patrone, "A possible smart metering system evolution for rural and remote areas employing unmanned aerial vehicles and internet of things in smart grids," *S*ensors, vol. 21, no. 5, p. 1627, 2021.
33. M. A. Ullah, K. Mikhaylov, and H. Alves, "Enabling mmtc in remote areas: Lorawan and leo satellite integration for offshore wind farm monitoring," *I*EEE Transactions on Industrial Informatics, vol. 18, no. 6, pp. 3744–3753, 2021.
34. V. S. Chippalkatti *e*t al., "Review of satellite based internet of things and applications," *T*urkish Journal of Computer and Mathematics Education (TURCOMAT), vol. 12, no. 12, pp. 758–766, 2021.
35. M. S. Abdalzaher, M. S. Soliman, S. M. El-Hady, A. Benslimane, and M. Elwekeil, "A deep learning model for earthquake parameters observation in iot system-based earthquake early warning," *I*EEE Internet of Things Journal, vol. 9, no. 11, pp. 8412–8424, 2021.
36. S.-W. Chen, J.-Q. Shao, and H.-S. Zhu, "Technology of internet of things responding to natural disasters," *E*AI Endorsed Transactions on Internet of Things, vol. 7, no. 26, pp. e2–e2, 2021.
37. A. S. Santos, A. C. Corsi, I. C. Teixeira, V. L. Gava, F. A. Falcetta, E. S. de Macedo, C. d. S. Azevedo, K. T. de Lima, and K. R. Braghetto, "Brazilian natural disasters integrated into cyber-physical systems: computational challenges for landslides and floods in urban ecosystems," in *2*020 IEEE International Smart Cities Conference (ISC2), pp. 1–8, IEEE, 2020.
38. G. Callebaut, G. Leenders, J. Van Mulders, G. Ottoy, L. De Strycker, and L. Van der Perre, "The art of designing remote iot devices-technologies and strategies for a long battery life," *S*ensors, vol. 21, no. 3, 2021.
39. M. Luglio, M. Marchese, F. Patrone, C. Roseti, and F. Zampognaro, "Performance evaluation of a satellite communication-based mec architecture for iot applications," *I*EEE Transactions on Aerospace and Electronic Systems, vol. 58, no. 5, pp. 3775–3785, 2022.
40. M. Pons, E. Valenzuela, B. Rodríguez, J. A. Nolazco-Flores, and C. Del-Valle-Soto, "Utilization of 5g technologies in iot applications: Current limitations by interference and network optimization difficulties-a review," *S*ensors, vol. 23, no. 8, p. 3876, 2023.
41. H. Allioui and Y. Mourdi, "Exploring the full potentials of iot for better financial growth and stability: A comprehensive survey," *S*ensors, vol. 23, no. 19, p. 8015, 2023.

42. V. Sharma, I. You, K. Andersson, F. Palmieri, M. H. Rehmani, and J. Lim, "Security, privacy and trust for smart mobile-internet of things (m-iot): A survey," *IEEE access*, vol. 8, pp. 167123–167163, 2020.
43. A. Tewari and B. B. Gupta, "Security, privacy and trust of different layers in internet-of-things (iots) framework," *Future generation computer systems*, vol. 108, pp. 909–920, 2020.
44. S. Khisa and S. Moh, "Medium access control protocols for the internet of things based on unmanned aerial vehicles: a comparative survey," *Sensors*, vol. 20, no. 19, p. 5586, 2020.
45. J. Bas and A. A. Dowhuszko, "On the use of nb-iot over geo satellite systems with time-packed optical feeder links for over-the-air firmware/software updates of machine-type terminals," *Sensors*, vol. 21, no. 12, p. 3952, 2021.
46. "The cropx system." Available at https://cropx.com/resources/ Date Last Accessed 02-January-2024.
47. J. Müller, O. Mitesser, H. M. Schaefer, S. Seibold, A. Busse, P. Kriegel, D. Rabl, R. Gelis, A. Arteaga, J. Freile, *et al.*, "Soundscapes and deep learning enable tracking biodiversity recovery in tropical forests," *Nature communications*, vol. 14, no. 1, p. 6191, 2023.
48. "Fleetcomplete." Available at https://blog.fleetcomplete.com/whats-new-december-2023/, Date Last Accessed 06-January-2024.
49. "Bp - pipeline integrity monitoring." Available at https://www.bp.com/, Date Last Accessed 28-January-2024.
50. "International committee of the red cross." Availabe at https://www.icrc.org/en, Date Last Accessed 10-January-2024.
51. Y.-H. Lee and F. J. Lin, "Enabling iot/m2m system scalability with fog computing," in *IoT as a Service: 5th EAI International Conference, IoTaaS 2019, Xi'an, China, November 16-17, 2019, Proceedings 5*, pp. 489–502, Springer, 2020.

Developing a Software-Defined Networking-Based Simulation Framework for the Internet of Space Things

Awais Aziz Shah and Muhammad Ali Jamshed

Abstract

Over the past decade, there has been a noticeable increase in the integration of smart devices, sensors, autonomous vehicles, drones, and robots into everyday human life, all falling under the broader umbrella of the Internet of Things (IoT). The next generation of IoT devices is anticipated to demand enhanced connectivity, minimal latency, and improved availability to uphold Quality of Service (QoS), while concurrently emphasizing energy efficiency. Presently, 4G/5G wireless and optical transport networks are utilized for existing IoT implementations. However, remote areas such as animal farms, agricultural lands, forests, and oceans encounter challenges related to insufficient or absent network connectivity. Although satellite communications aim to address these issues, the significant costs associated with this approach present a substantial challenge. Moreover, the existing fleet of satellites may not be adequately equipped to accommodate the growing number of IoT deployments in the near future. In response to these challenges, the concept of the Internet of Space Things (IoST) has emerged as a transformative solution. IoST facilitates the deployment of small-sized satellites in Low Earth Orbit (LEO), enabling direct communication through space and providing an enhanced QoS. This development is expected to overcome existing obstacles in remote areas. This

A. A. Shah
School of Computing, University of Glasgow, Glasgow, UK

M. A. Jamshed (✉)
College of Science and Engineering, University of Glasgow, Glasgow, UK
e-mail: muhammadali.jamshed@glasgow.ac.uk

© The Author(s), under exclusive license to Springer Nature Switzerland AG 2025
M. A. Jamshed and A. Nauman (eds.), *Integrated Terrestrial and Non-Terrestrial Networks*, Synthesis Lectures on Communications,
https://doi.org/10.1007/978-3-031-63705-6_3

chapter introduces a simulation framework grounded in Software-Defined Networking (SDN) to orchestrate a large-scale satellite network. The framework utilizes a multi-level SDN controller and a centralized orchestrator to efficiently manage the deployment and operation of satellites.

Keywords

Software-Defined Networking (SDN) • Internet of Space Things (IoST) • Interent of Things (IoT) • Low Earth Orbit (LEO) satellites • Quality of Service (QoS)

1 Introduction

With the rapid advancements in telecommunications and the implementation of Low-Power Wide-Area Networking (LPWAN) technologies, the global landscape is now saturated with a multitude of Internet of Things (IoT) devices [1]. In the last decade, there has been a rapid increase in the integration of smart devices across various applications, and it is estimated that by the onset of 2022, over 20 billion IoT devices were operational worldwide [2]. Despite this widespread adoption, a significant challenge has arisen in remote areas characterized by limited or no connectivity, along with the associated high expenses of establishing such connectivity [3].

To address the issue of remote connectivity, a temporary solution within the 5G paradigm has emerged, referred to as the PopUP network [4]. This network is designed to offer provisional connectivity in field environments. However, the deployment costs associated with this technology are prohibitively high, rendering it impractical for small and medium-scale use cases.

The global demand for Satellite communication within the IoT network has led to the recognition of satellite solutions as crucial for achieving extensive coverage, especially in remote areas and forests. However, the conventional approach involves the development, deployment, and launch of large-scale satellites into space, incurring significant costs in the billions of dollars. In contrast, a recent development in satellite technology is the emergence of small, cost-effective satellites known as 'CubeSats' [5]. These CubeSats, primarily utilized by universities for research missions in the Low Earth Orbit (LEO), serve as a promising avenue for future telecommunications and play a key role in enabling the Internet of Space Things (IoST) (IoST).

CubeSats are remarkably compact, with dimensions of approximately 10 cm × 10 cm × 10 cm, roughly equivalent to the size of a Rubik's cube, and weigh around 1 Kg. Statistics indicate that over 1,600 CubeSats have been launched into orbit, with approximately 90 experiencing destruction during launch failures. Apart from their cost-effectiveness, CubeSats offer the flexibility of being deployed individually or stacked together, allowing for the combination of up to 24 units.

In spite of the benefits associated with cost-effectiveness and flexibility, the extensive incorporation of CubeSats into industrial settings presents a formidable challenge. Various complexities contribute to this challenge, encompassing issues such as the centralized orchestration and administration of a substantial quantity of CubeSats, the deployment of Virtual Network Functions (VNF), scheduling, application deployment, and the formulation of communication protocols in the space environment [6].

Software-Defined Networking (SDN) emerges as a groundbreaking technology facilitating the virtualization of networking resources and their centralized management through an entity known as the Orchestrator. This study introduces a simulation framework based on SDN to orchestrate a significant number of satellites (CubeSats) within the IoST paradigm. The framework incorporates multiple tiers of SDN controllers, including a level-1 SDN controller situated on ground stations, providing information about communicating satellites to the level-2 SDN controllers. Additionally, a centralized orchestrator gathers global information from the level-2 controller, making decisions to optimize the overall satellite network and deploy VNF within the network.

The remainder of this chapter is structured as follows: Sect. 2 delivers an overview of the current landscape of emerging satellite technologies and the IoST. Section 3 performs a survey of potential use cases for CubeSats technology. The development of the suggested emulation framework employing SDN is outlined in Sect. 4. Finally, Sect. 5 serves as the concluding segment for this chapter.

2 Literature Review

Traditional satellites face various drawbacks, encompassing, but not confined to:

- **High costs**: The cost of a conventional satellite is generally estimated to range from 150 to 350 million dollars, whereas a CubeSat could be acquired for less than 200,000 dollars [7].
- **Exposure to high risks**: A breakdown at any point during the satellite development process can result in significant financial losses and time setbacks.
- **Congestion risk**: Traditional satellites utilize widely used frequency bands, which are increasingly susceptible to congestion in contemporary times [8]. Additionally, incorporating different frequencies into existing satellites would pose significant challenges.
- **Extended duration**: The creation of a satellite could span a duration of seven years or longer, particularly in the case of government programs [9].

The identified limitations have led to increased exploration of alternative, more affordable satellite options, including microsatellites like CubeSats. Figure 1 illustrates the dimensions of a standard CubeSat, while Fig. 2 depicts a stacked arrangement of 24 CubeSats.

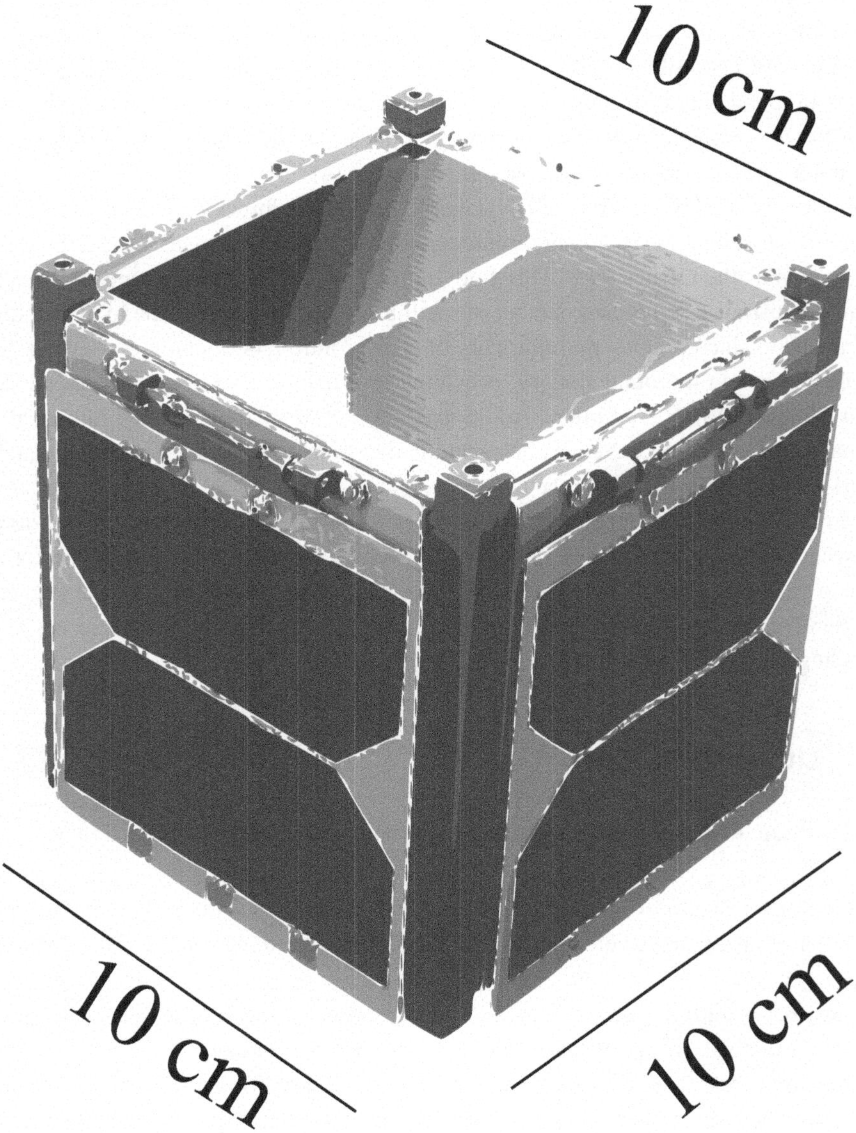

Fig. 1 An illustration of a single CubeSat

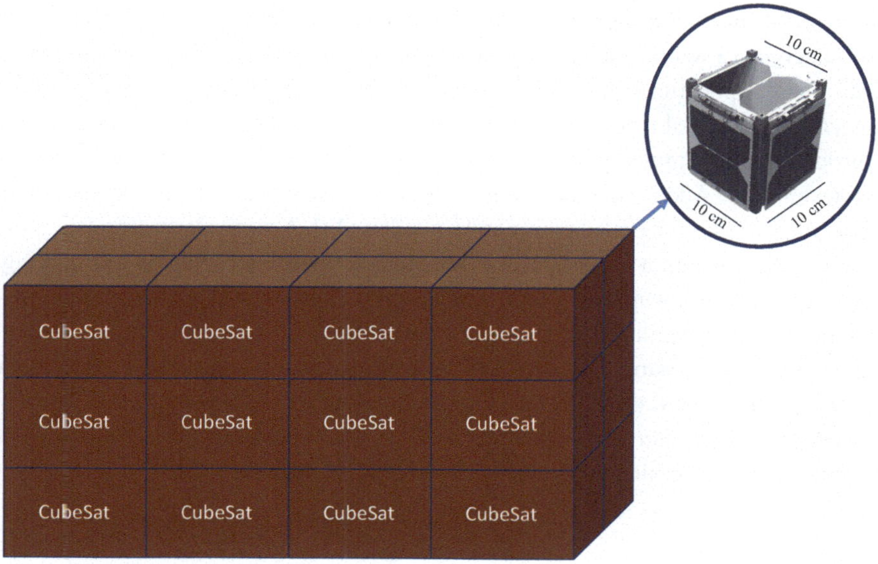

Fig. 2 An illustration of stack of 24 CubeSats

Review articles addressing CubeSats began to emerge in 2017, encompassing proposals for novel antenna designs, advancements in small satellite technology, and investigations into CubeSats' mission objectives and communication protocols. Further exploration [10] delves into a comparative analysis of various CubeSats communication system architectures, with a focus on evaluating performance metrics such as power consumption, data rate, and error rate. A significant portion of the scrutinized literature relies on Software Defined Radios (SDR) and Field Programmable Gate Arrays (FPGA). While SDR offer cost-effectiveness and compact sizes, they fall short in data rate compared to commercial hardware. Despite the increased complexity and development time associated with SDR, their inherent scalability serves as a primary motivator for systems like the AeroCube SDR. This system aims to facilitate upgrades with more spectrally efficient waveforms and serves as an in-orbit testing platform for pioneering communication techniques before their integration into newer AeroCube satellite iterations [10].

Furthermore, research has been conducted to assess the performance of SDN-based IoST [11]. The proposed framework is structured around three essential domains: system infrastructure, control and management (C&M), and policy and orchestration (P&O). Evaluation criteria include Ground-Satellite Links (GSL), Inter-Satellite Links (ISL), next-hop metrics, and the overall operation of the system. The GSL assesses the highest attainable data rate across the IoST Hub-CubeSat Link, and next-hop metrics offer valuable insights into network topology, enriching perspectives in analytical modeling.

Extensive research has been conducted in the realms of image processing and image segmentation to comprehend the capabilities of CubeSats.

- **Terrestrial thermal imaging** [12]: The landscape of scientific observation applications has undergone a significant transformation with the introduction of remote sensing CubeSats. These applications span diverse areas, including terrestrial thermal imaging, weather forecasting, and real-time cloud monitoring. The envisioned system takes into careful consideration the limited power resources available in missions employing microsatellite technology. Addressing the energy aspect, extensive research has explored various battery options for integrated power and CubeSats [13]. A particularly promising avenue involves the utilization of advanced nanomaterials, specifically Silicon alloys designed for Li-ion batteries, which may represent the most effective strategy for maintaining the viability of microsatellites in the space environment.
- **Space image processing** [14]: CubeSats find utility in extracting Line-of-Sight information for planets. The suggested investigation demonstrates the effectiveness of employing the RANSAC algorithm and calculating the covariance matrix for projecting the position of the planet, showcasing their robustness in this application.

2.1 CubeSats

The concept of CubeSats was initially introduced in 2001 as compact picosatellites designed for educational and research purposes [15]. Originating from a program initiated in 1999 at Stanford University, CubeSats are constructed using one or more 10 cm $\times$ 10 cm $\times$ 10 cm cubes, each weighing approximately 1 kg. The utilization of commercial off-the-shelf hardware components allows for cost-effectiveness and accessibility to researchers [7].

Due to their modular nature, CubeSats offer flexibility in combination, enabling the adjustment of the final product's maximum size and weight according to specific objectives. In contemporary applications, CubeSats are not only utilized for image recognition but also for onboard machine learning-based applications.

CubeSats present also some limitations:

- **Data storage**:
  The standard memory capacity of CubeSats generally falls within the range of 32 kB to 8 MB, with the potential for expansion to a total memory of 8 GB through the inclusion of supplementary flash memories [16]. However, this memory capacity may not suffice for storing numerous multispectral images, especially in applications related to aerial imagery.
- **Communications**:
  Research on CubeSat communication underscores the necessity for any antenna design to conform to size and weight restrictions while still achieving optimal antenna radiation performance [17]. A significant number of CubeSats typically operate in the UHF band, offering a maximum data rate of 9600 bps. In contrast, a minority (15% of the surveyed

CubeSats) use the S-band, which provides a higher maximum data rate of 256 kbps [18]. The constrained data rates are primarily dictated by the link power budget's availability rather than the absence of suitable electronics [18].

- **Power consumption**:
  A CubeSat typically has an available power of approximately 1W [19]. While solar panels can be utilized to support a CubeSat, the overall power capacity may prove inadequate for specific applications. This limitation stems from the inadequacy of these power levels to support high-energy instruments like lidars and radar altimeters [19].

3 IoST: Use Cases

Microsatellites hold promise for addressing a multitude of applications in the near future. Apart from the aforementioned roles in image processing, below are several other potential use cases outlined:

- **Global freight mobility**:
  Efficient communication plays a pivotal role in the global transportation infrastructure. With the increasing demand for global freight mobility, the IoST emerges as a cost-effective solution for monitoring and facilitating communication in global freight mobility. This is especially relevant in scenarios such as the navigation of large cargo ships through sea channels. Presently, owners of cargo ships bear significant subscription fees to satellite companies for communication services. CubeSats can contribute to mitigating these elevated subscription costs.
- **Internet-of-Drones**:
  Anticipations suggest that drones will soon have diverse applications, including delivering goods, traffic monitoring, and emergency services. With CubeSats operating in LEO, there is the potential for them to optimize the coordination and communication of drones within the Internet of Drones paradigm [20].
- **Archaeology and forestry**:
  CubeSats possess the potential to serve archaeologists by monitoring changes in the Earth's landscape. They can also be valuable for forestry departments, aiding in the surveillance of tree cutting and other events in extensive forests that are impractical to cover manually. Furthermore, the capability to analyze top-view images of forests captured by CubeSats can enhance efforts to prevent the spread of fires [21].
- **Smart home devices**:
  While IoT-enabled smart home devices have experienced swift adoption in urban areas worldwide, individuals residing in remote areas, deserts, and mountainous regions with limited or no network connectivity are unable to enjoy the benefits of these technologies. The implementation of the IoST is poised to enable the integration of IoT in such areas.

- **Autonomous vehicles**:
  In the past decade, autonomous vehicles have been introduced to the market, but their widespread adoption remains limited to a few developed cities and countries globally. A significant impediment to their adoption in developing countries is the challenge of communication issues. The maturation of the IoST paradigm based on CubeSats holds the promise of resolving these communication challenges, thereby opening the doors for the global adoption of autonomous vehicles.
- **Block-chain and agriculture**:
  Smart Agriculture is an evolving technology that integrates Information Technology, satellite technology, the IoT, and geographical information systems. In remote farming regions, issues emerge related to network connectivity, making the deployment of IoT devices a challenging endeavor. The IoST provides a remedy by enabling the implementation of smart agriculture, enabling farmers to harness emerging blockchain technologies within the agricultural domain.

4 SDN-Based Framework for IoST

In the coming years, an abundance of CubeSats or similar satellites based on the IoST is expected to be deployed in LEO on a global scale. An anticipated challenge will be centered around the centralized management, deployment of VNF, and orchestration of these satellites. To tackle these challenges, this chapter introduces a simulation framework based on SDN with multiple levels for the centralized management of satellites. Beyond assisting in management tasks, the framework aims to virtualize the network infrastructure of satellite systems, with the goal of improving Quality of Service (QoS), reducing latency, ensuring high availability, and optimizing user experience. The proposed framework, illustrated in Fig. 3, consists of a Level-1 SDN controller, multiple SDN controllers situated at Level-2 on ground stations, and a centralized orchestration layer. Additional details are provided below.

4.1 Ground Station SDN Controller: Level-2

Ground stations host these SDN controllers, facilitating communication with multiple satellites within their coverage area. The data gathered from various satellites is then relayed to the level-2 SDN controller.

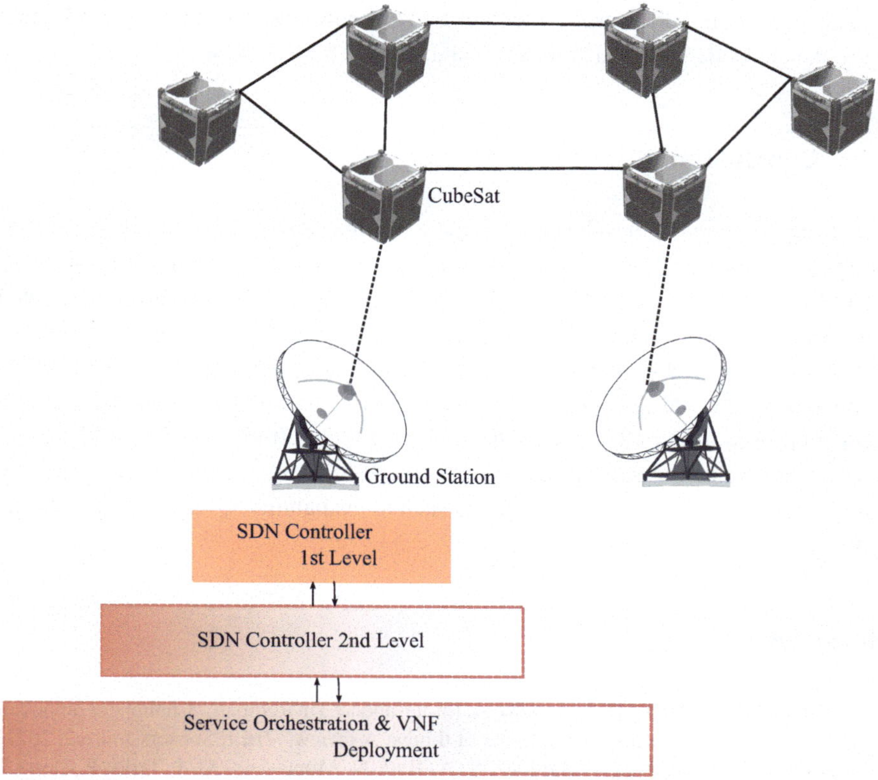

Fig. 3 Conceptual framework for CubeSat emulation utilizing SDN technology

4.2 SDN Controller: Level-1

The specified SDN controller tier gathers data on satellite positioning, communication status, and network topology to establish a centralized perspective. Furthermore, it directs the level-1 SDN controllers to transmit instructions to the satellites for deploying VNF and improving the overall satellite network.

4.3 Orchestration Layer: Level-3

The pivotal function of the centralized orchestrator lies in supervising the comprehensive status of the satellite network and formulating decisions to guarantee its smooth functionality. Its duties encompass the improvement of QoS, user experience enhancement, latency reduction, and the optimization of global availability for users. To attain these objectives, the orchestrator issues directives to the level-1 SDN controller, which in turn communicates them to the controllers located at local ground stations for execution.

5 Conclusion

In this chapter, we have introduced a simulation framework specifically designed for the centralized supervision and coordination of ‘cubeSats’ (Satellites) within the context of the IoST paradigm. The framework comprises three tiers: level-1 SDN controllers situated on ground stations, which relay information to the level-2 SDN controller. The centralized orchestrator is linked with the level-2 controller to convey requirements for the effective administration of the infrastructure. The goal is to ensure optimal user experience, minimal latency, improved availability, and maintain QoS. Future efforts will focus on the implementation and evaluation of the proposed framework, including a comparative analysis with existing satellite technologies across factors such as availability, QoS, user experience, and latency.

References

1. M. A. Jamshed, K. Ali, Q. H. Abbasi, M. A. Imran, and M. Ur-Rehman, “Challenges, applications and future of wireless sensors in internet of things: A review,” *IEEE Sensors Journal*, 2022.
2. M. Naeem, T. Jamal, J. Diaz-Martinez, S. A. Butt, N. Montesano, M. I. Tariq, E. De-la Hoz-Franco, and E. De-La-Hoz-Valdiris, “Trends and future perspective challenges in big data,” in *Advances in Intelligent Data Analysis and Applications*, pp. 309–325, Springer, 2022.
3. N. Islam, M. M. Rashid, F. Pasandideh, B. Ray, S. Moore, and R. Kadel, “A review of applications and communication technologies for internet of things (iot) and unmanned aerial vehicle (uav) based sustainable smart farming,” *Sustainability*, vol. 13, no. 4, p. 1821, 2021.
4. R. Trivisonno, R. Guerzoni, I. Vaishnavi, and D. Soldani, “Sdn-based 5g mobile networks: architecture, functions, procedures and backward compatibility,” *Transactions on Emerging Telecommunications Technologies*, vol. 26, no. 1, pp. 82–92, 2015.
5. I. F. Akyildiz and A. Kak, “The internet of space things/cubesats,” *IEEE Network*, vol. 33, no. 5, pp. 212–218, 2019.
6. A. A. Shah, “A software-defined networking based simulation framework for internet of space things,” in 2023 IEEE 97th Vehicular Technology Conference (VTC2023-Spring), pp. 1–4, IEEE, 2023.
7. F. Davoli, C. Kourogiorgas, M. Marchese, A. Panagopoulos, and F. Patrone, “Small satellites and cubesats: Survey of structures, architectures, and protocols,” *International Journal of Satellite Communications and Networking*, vol. 37, no. 4, pp. 343–359, 2019.
8. S. Madry, *Space systems for disaster warning, response, and recovery*. Springer, 2015.
9. L. A. Davis and L. Filip, “How long does it take to develop and launch government satellite systems,” *International Cost Estimating and Analysis Association” AEROSPACE REPORT*, 2015.
10. A. Zeedan and T. Khattab, “A critical review of baseband architectures for cubesats communication systems,” *arXiv preprint arXiv:2201.09748*, 2022.

11. A. Kak, E. Guven, U. E. Ergin, and I. F. Akyildiz, "Performance evaluation of sdn-based internet of space things," in 2018 IEEE Globecom Workshops (GC Wkshps), pp. 1–6, IEEE, 2018.
12. C. Nagarajan, R. G. D'souza, S. Karumuri, and K. Kinger, "Design of a cubesat computer architecture using cots hardware for terrestrial thermal imaging," in 2014 IEEE International Conference on Aerospace Electronics and Remote Sensing Technology, pp. 67–76, IEEE, 2014.
13. A. F. Hepp, P. N. Kumta, O. I. Velikokhatnyi, and R. P. Raffaelle, "Batteries for integrated power and cubesats: Recent developments and future prospects," Silicon Anode Systems for Lithium-Ion Batteries, pp. 457–508, 2022.
14. E. Andreis, P. Panicucci, V. Franzese, and F. Topputo, "A robust image processing pipeline for planets line-of-sign extraction for deep-space autonomous cubesats navigation," in 44th AAS Guidance, Navigation and Control Conference, pp. 1–19, 2022.
15. J. Puig-Suari, C. Turner, and R. Twiggs, "Cubesat: the development and launch support infrastructure for eighteen different satellite customers on one launch," 2001.
16. A. Toorian, K. Diaz, and S. Lee, "The cubesat approach to space access," in 2008 IEEE aerospace conference, pp. 1–14, IEEE, 2008.
17. S. Abulgasem, F. Tubbal, R. Raad, P. I. Theoharis, S. Lu, and S. Iranmanesh, "Antenna designs for cubesats: A review," IEEE Access, vol. 9, pp. 45289–45324, 2021.
18. J. Bouwmeester and J. Guo, "Survey of worldwide pico-and nanosatellite missions, distributions and subsystem technology," Acta Astronautica, vol. 67, no. 7-8, pp. 854–862, 2010.
19. D. Selva and D. Krejci, "A survey and assessment of the capabilities of cubesats for earth observation," Acta Astronautica, vol. 74, pp. 50–68, 2012.
20. M. A. Jamshed, A. Kaushik, M. Toka, W. Shin, M. Z. Shakir, S. P. Dash, and D. Dardari, "Synergizing airborne non-terrestrial networks and reconfigurable intelligent surfaces-aided 6g iot," arXiv preprint arXiv:2310.16371, 2023.
21. M. A. Jamshed, C. Theodorou, T. Kalsoom, N. Anjum, Q. H. Abbasi, and M. Ur-Rehman, "Intelligent computing based forecasting of deforestation using fire alerts: A deep learning approach," Physical Communication, vol. 55, p. 101941, 2022.

Software-Defined Space-Air-Ground IoT: Motivation, Architecture and Research Challenges

Jehad Ali, Gaoyang Shan, Ahmed Raza Mohsin, Maira Khalid, Jisi Chandroth, and Byeong-hee Roh

Abstract

The Software-Defined Space-Air-Ground Internet of Things (IoT) (SDSAG-IoT) is an emerging paradigm that integrates various types of wireless networks, including satellite, air, and ground networks. This Chapter provides an architecture for SDSAG-IoT with an explanation of SDN layers and the organization of the control plane. Moreover, we also discuss the role of Artificial Intelligence (AI) in improving network automation, predictive analytics, security, and performance. The Chapter also discusses the challenges, use of reinforcement learning, and future research directions for SDSAG-IoT, including the need for massive amounts of data, robust and secure data processing and storage, and the development of more sophisticated algorithms for network optimization, security, and analytics. The study concludes that SDN and AI have the potential to revolutionize Space-Air-Ground IoT, but organizations must address the challenges and continue to invest in research and development to fully realize their potential.

Keywords

Software-defined networking • IoT • Space-Air-Ground IoT • Artificial intelligence • 6G • Reinforcement learning

J. Ali (✉) · A. R. Mohsin · M. Khalid · J. Chandroth · B. Roh
Department of AI Convergence Network, Ajou University, Suwon, South Korea
e-mail: jehadali@ajou.ac.kr

G. Shan
Department Software and Computer Engineering, Ajou University, Suwon 16499, South Korea

© The Author(s), under exclusive license to Springer Nature Switzerland AG 2025
M. A. Jamshed and A. Nauman (eds.), *Integrated Terrestrial and Non-Terrestrial Networks*,
Synthesis Lectures on Communications,
https://doi.org/10.1007/978-3-031-63705-6_4

1 Introduction

Space air-to-ground (SAG) infrastructure refers to the network infrastructure that enables communication and data transfer between space-based assets, airborne vehicles, and ground-based devices. This infrastructure plays a critical role in many applications, such as satellite communication, remote sensing, Internet-of-Things (IoT), and unmanned aerial vehicle (UAV) operations [1]. Software-defined networking (SDN) is an approach to computer networking that allows for greater flexibility, agility, and programmability than traditional networking architectures. By separating the control plane from the data plane, SDN enables network administrators to manage and configure network resources more dynamically, in a programmable way, and to respond more quickly and accurately to changing network conditions [2, 3].

Moreover, One of the key advantages of SDN in SAG infrastructure is its ability to dynamically allocate network resources, such as bandwidth and routing paths, based on real-time network conditions monitoring. This allows for efficient data transfer and reduces latency and delay, which is particularly important in applications such as UAV operations, IoT, Industrial IoT [4], vehicular networks [5], 6G [6, 7], and remote sensing. Similarly, when it comes to SAG-IoT and Vehicular networks, SDN has the potential to play a critical role in enabling these complex, distributed networks to operate more effectively and efficiently. By using SDN to centrally manage and control network resources, organizations can achieve greater visibility into network traffic due to its global view of the network, optimize network performance, and more effectively manage network security.

In addition, one of the key benefits of SDN in these contexts is its ability to enable network slicing. Network slicing allows for the creation of isolated, virtual networks leveraging network function virtualization (NFV) [8] within a larger physical network, each with its own set of resources and policies. This is particularly useful in SAG-IoT, where different types of traffic may have different requirements in terms of bandwidth, latency, and security. Furthermore, the SDN can also be used to enable more dynamic and flexible routing in these networks. By using programmable network switches and routers, network administrators can more easily adapt to changing network conditions and traffic patterns and can adjust routing policies in real time to optimize network performance. Consequently, the use of SDN in SAG-IoT and Vehicular networks has the potential to enable more efficient, flexible, and secure networking, and to help organizations achieve their goals more effectively in these complex and challenging environments [9].

The SDSAG-IoT is a network of interconnected devices, sensors, and machines that can communicate with each other over a software-defined network. This network is comprised of a variety of devices, including those in space, the air, and on the ground. The space component of this network includes satellites, which can be used to collect and transmit data to other devices on the network. The air component includes drones and other unmanned aerial vehicles (UAVs), which can be used for surveillance, monitoring, and other purposes. The ground component includes a variety of devices, such as sensors, cameras, and other equipment, that can be used to gather and transmit data [10].

The software-defined aspect of this network refers to the fact that the network is controlled by software, rather than hardware. This means that devices on the network can be reconfigured and reprogrammed as needed, allowing for greater flexibility and adaptability. In a nutshell, the SDSAG-IoT has the potential to revolutionize a wide range of industries, from transportation and logistics to agriculture and environmental monitoring. By enabling devices to communicate and work together in real time, this network can help organizations make more informed decisions and improve operational efficiency.

The growth of wireless networks, from 2 to 4G and beyond, has consistently changed information services to fulfil varied societal and vertical field needs. 5G brought scenarios such as eMBB, uRLLC, and mMTC, which cater to varying data rate, latency, and machine-type communications requirements. In the future, intelligence, complexity, and customization will be distinguishing characteristics of an Information-Communication-Data convergence network. 6G aspires to provide an assortment of intelligent applications, requiring ubiquitous worldwide coverage and high-speed access, spurring the development of Space-Air-Ground Integrated Networks (SAGINs). These networks promise different services across wide geographical areas but present challenges in organizing resources and managing complex movement patterns. SAGINs, which combine satellites, High-Altitude Platforms (HAPs), unmanned aerial vehicles (UAVs), and terrestrial networks, provide seamless worldwide coverage critical for a variety of applications in the 6G era. Satellites and HAPs improve coverage and internet availability, whilst UAVs allow quick and flexible responses in situations such as events or places with poor terrestrial network coverage. However, because SAGINs cover many domains, dynamic situations, and various resources, providing on-demand services poses obstacles, necessitating detailed architectural and technology research for optimal service fulfilment in this environment. This viewpoint on 6G service-oriented SAGINs emphasizes the importance of thorough research that addresses the challenges of dynamic network settings and multi-dimensional resource utilization [11].

In the context of Space-Air-Ground Integrated Networks (SAGINs), which are crucial to the 6G era, the combination of satellites, High-Altitude Platforms (HAPs), UAVs, and terrestrial networks provides ubiquitous worldwide coverage, which is critical for a variety of applications. Satellites and HAPs increase coverage and internet availability, whilst UAVs provide quick, flexible responses in instances with insufficient terrestrial network coverage, such as events. However, the complexity of SAGINs, which span many domains, dynamic situations, and diverse resources, offers obstacles in offering on-demand services. For efficient service fulfilment within this complex environment, much study in architecture and technology is required. To address these challenges, a fresh viewpoint focuses on using queuing game theory to describe SDSAGIN transmission control, optimizing techniques to maximize the system's social welfare. The method takes into account the various user categories and their different demands inside SAGINs, making the network model more practical. The addition of observable and unobservable queuing modes for the controller node improves information transmission efficiency while decreasing interactions in SAGINs due to excessive delay and jitter. The proposed DHSTC scheme improves system performance by minimizing interaction and determining the best strategy based on whether the controller's queuing situation is visible to the switch node, significantly contributing to adaptive system capabilities for managing switch node entries and exits. [12].

The Internet of Vehicles (IoV) has transformed the automotive industry by enabling vehicular networks (V2V and V2I), improving perception and safety with services such as traffic information and remote diagnosis, and catering to infotainment needs through file transfers and online games. IoV, on the other hand, confronts difficulties in dynamic network administration, mobility, and load balancing, pushing the development of software-defined networking (SDN) as a possible solution. Despite the benefits of SDN, security flaws in SDN-controlled vehicular networks pose threats such as location hijacking attacks, which could jeopardize car services and user safety. To address this, research focuses on identifying attacks but falls short of providing adequate recovery methods. This research is the first to investigate location hijacking attacks on SDN controllers and presents a deep Q-learning (DQL)-based recovery mechanism that ensures service continuity by selecting relocating impacted services to High Altitude Platforms (HAPs) inside a space-air-ground-integrated vehicular network. This work not only emphasizes the vulnerability of SDN in vehicular networks, but also proposes a recovery mechanism to limit the impact of location hijacking attacks, safeguarding vehicle services within an integrated network infrastructure [13].

In the area of IoT systems, the convergence of cloud computing, next-generation mobile communication (5G and beyond), and the rise of edge computing have opened the way for Mobile Edge Computing (MEC) to cater to latency-sensitive applications such as mobile gaming and augmented reality. However, implementing efficient compute offloading solutions for various IoT scenarios is difficult. While terrestrial-based MEC research concentrates on minimizing task execution delay and energy consumption, it fails to provide robust performance in remote places where terrestrial infrastructure is unavailable. To fill this void, the introduction of Space-Air-Ground Integrated Networks (SAGINs) promises ubiquitous compute services in remote places lacking terrestrial infrastructure. Edge Computing-enabled Space-Air-Ground Integrated Networks (EC-SAGIN) in particular use UAVs and satellites to provide flexible computing and communication services. Despite substantial research in EC-SAGIN on task scheduling/offloading, developing methods for power-constrained IoT devices, UAVs, and satellites remains difficult. This Chapter offers an energy-efficient EC-SAGIN system in which power-constrained LEO satellites, UAVs, and IoT devices collaborate to perform computing tasks. The dynamic processing and communication capabilities of satellites and UAVs necessitate an adaptive task offloading technique, which is addressed inside EC-SAGIN by a Deep Reinforcement Learning (DRL)-based algorithm that takes uncertainties and dynamics into account.

Contributions include establishing an energy-efficient offloading method, framing an optimization problem for energy-saving task processing, and constructing a DRL algorithm for adaptive decision-making in limited EC-SAGIN contexts [14].

Beyond 5G (B5G) and 6G communication technologies are aiming for high-speed, low-latency, and high-density capabilities, with considerable improvements over 5G. The 6G requirements, which include peak download rates of 1 Tbit/s, ultra-low transmission delays, and device density of 100/m3, predict seamless coverage throughout 99% of terrestrial and marine areas via a Space-Air-Ground Integrated Network (SAGIN). SAGIN, which combines satellites, drones, and ground networks, poses complicated difficulties such as resource allocation, security, and traffic management. Using AI and SDN in SAGIN applications aids in adaptive policy adjustments, improving network stability, Quality of Service (QoS), and Quality of Experience (QoE). To address the complexity, a hybrid hierarchical control approach combining distributed and centralized methods is being considered, along with Deep Reinforcement Learning (DRL) techniques to optimize network control, resource allocation, and service orchestration, offering a potential solution to SAGIN's intricate challenges. This hierarchical hybrid DRL design uses local and global controllers to break complex policies into simpler subgoals, efficiently optimizing SAGIN in its diverse environment [15].

Advances in science and technology are paving the way for the Internet of Things (IoT) to become the backbone of future networks, expanding spatially and in communication content. While 5G networks support different applications such as virtual reality and smart cities, they struggle to cover far places such as oceans and space, necessitating the need for larger communication dimensions. SAGINs (Space-Air-Ground Integrated Networks) emerge as a solution, connecting space, air, and ground networks to deliver pervasive and collaborative information infrastructure across huge spaces. SAGIN architecture, which includes ground, air, and space-based networks, strives for broad coverage, network integration, and a game-changing information revolution. However, administering SAGINs via centralised or distributed architectures is difficult due to resource management, storage utilisation, dynamic request processing, and network topology support. To solve these issues, this work models SAGIN resource allocation as a Markov decision process (MDP) and proposes a distributed Deep Reinforcement Learning (DRL)-based resource management algorithm. It also introduces a distributed management architecture for space, air, and ground-based networks, and it proves superior performance through simulations and comparisons with existing resource management systems [16].

The rise of connected and automated vehicle (CAV) technology is poised to alter transportation networks through autonomous decision-making and communication among cars, promising increased safety, efficiency, and sustainability. However, because to spectrum scarcity and fixed infrastructure, contemporary terrestrial networks suffer difficulties in providing ubiquitous and dependable services. Although low-Earth orbit (LEO) satellite networks and agile aerial networks via unmanned aerial vehicles (UAVs) are potential answers, they also present problems such as propagation delay and limited durability. As a result, it is critical to merge these networks into a space-air-ground integrated vehicular network (SAGVN). Managing multi-dimensional resources across various segments for optimal quality of service (QoS) in SAGVN is an issue that has been solved here using a software-defined networking (SDN) architecture. Traditional model-based approaches are ineffective, necessitating the need for AI-driven solutions to efficiently manage the dynamic SAGVN environment. This Chapter proposes an SDN-based SAGVN framework that uses a hybrid control architecture and AI approaches to improve resource management via network slicing, mobility management, and intelligent content caching and delivery [17].

The main contributions of our Chapter are as follows.

- We propose a detailed architecture for SDSAG-IoT.
- Moreover, the article also provides a discussion on the use of AI and its advantages in the proposed architecture for SDSAG-IoT.
- In addition, the challenges, use of reinforcement learning, and future directions in the Software-Defined SAG-IoT are also highlighted.

The rest of our Chapter is organized as follows. In Sect. 2, we discuss the motivation for our proposed architecture. In Sect. 3, a detailed overview of the Software-Defined space-air-ground IoT is provided. Section 4 illustrates the challenges of the proposed architecture. The Chapter is concluded in Sect. 5 with future directions.

2 Motivation for Software-Defined Space Air Ground Iot

The SDN can be highly beneficial in SAG-IoT. As, SDN allows for greater flexibility and agility in the network, which is essential in a dynamic environment like SAG-IoT. With SDN, devices on the network can be reconfigured and reprogrammed in real-time to respond to changing conditions and requirements. This allows for greater adaptability and faster response times, which can be critical in applications like disaster management and response and surveillance.

In addition, SDN enables the creation of virtual networks leveraging NFV, and network slicing which can be used to segment the network into smaller, more manageable pieces. This is particularly useful in SAG-IoT, where there may be a large number of devices and sensors on the network. By segmenting the network into virtual networks, organizations can improve security and reduce the risk of unauthorized access. Moreover, another advantage of SDN in SAG-IoT is that it enables central management and orchestration of the network. This means that network policies and configurations can be managed from a central location with a global view, rather than having to be configured on individual devices in a distributed manner. This simplifies network management and reduces the risk of misconfigurations or errors.

Overall, SDN can play a crucial role in SAG-IoT by improving network agility, security, and management. By enabling devices to communicate and work together seamlessly, SDN can help organizations make more informed decisions and improve operational efficiency in a variety of industries.

3 Architecture of Software-Defined

SPACE-AIR-GROUND IOT The architecture of SDSAG-IoT is described as under.

The architecture of SDN for Space-Air Ground IoT typically includes three layers: the application layer, the control layer, and the infrastructure layer. In [3], the authors describe the architecture for vehicular networks. Herein, we extend the architecture to IoT. Our proposed architecture also gives an overview of the introduction of AI in the application layer as well as the benefits and utilization of reinforcement learning. Figure 1 explains the detailed proposed architecture consisting of the application, control, and IoT infrastructure layer.

Moreover, there is air and space infrastructure exchanging information and getting control policies from the hierarchical control layer of SDN. In addition, AI and reinforcement learning applications and algorithms can be deployed from the application layer.

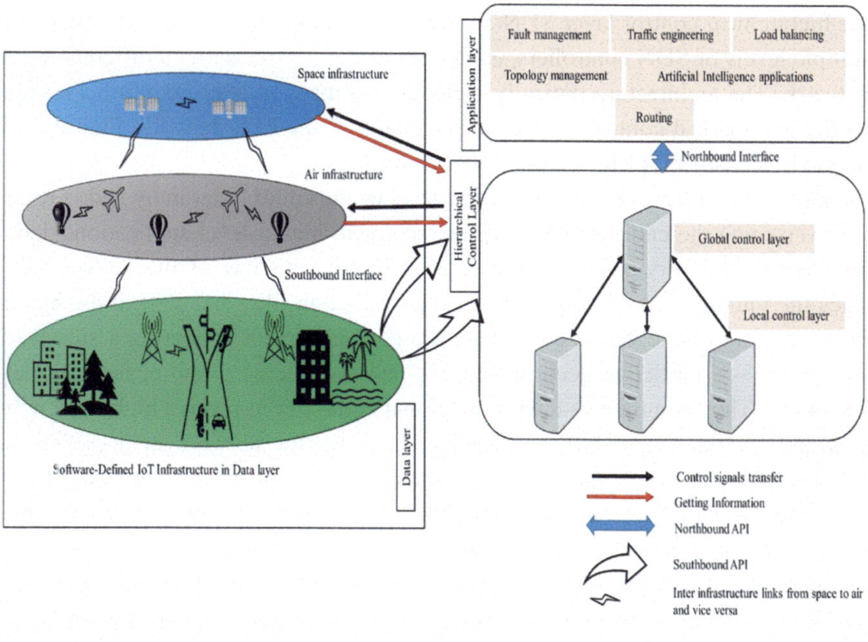

Fig. 1 Software Defined Space-Air-Ground IoT architecture

A. **Application Layer**

At the application layer, applications and services interact with the network. This layer includes applications that run on devices and sensors in the SAG-IoT network, as well as applications that run on external servers and cloud platforms enabled with cloud computing capabilities. Applications at this layer may include data analytics, machine learning, reinforcement learning, AI, and other services that help to extract insights from the data collected by the network.

B. **Control Layer**

The control layer is responsible for managing the network and communicating with the application layer. This layer includes the SDN controller, which is responsible for managing the flow of traffic through the network. The SDN controller communicates with the application layer to receive instructions on how to manage the network and then communicates with the infrastructure layer to implement those instructions. Different controllers are employed for managing the policies and control of the network in an end-to-end (E2E) manner such as OpenDayLight, ONOS, POX, Trema, Ryu, etc.

C. **Hierarchical Control Layer**

The hierarchical control layer SDN architecture is a network architecture that consists of multiple levels of SDN controllers, each responsible for managing a different aspect of the network. The architecture typically includes two main layers of controllers. Figure 2 shows the hierarchical controller layer management of the E2E network. Figure 2 shows the hierarchy consisting of local and global controllers.

The global controller: This is the highest level of the control hierarchy and is responsible for overseeing the entire network. It interacts with higher-level applications deployed as applications and services which provides a high-level view of the network. It also manages the middle layer of controllers or the local controllers below it in the hierarchy.

The local controllers: These controllers are responsible for managing a subset of the network, such as a particular geographical region or a specific group of network devices (in a domain). They communicate with the global controller to receive high-level network policies and translate them into low-level policies for the network devices. They also provide feedback to the global controllers about the state of the network devices. They also manage the data layer IoT sensors [18–20]. These controllers are responsible for managing individual network devices, such as switches and routers.

The hierarchical control layer SDN architecture provides several benefits, such as scalability, fault tolerance, and efficient resource utilization. By dividing the network into multiple levels of controllers, the architecture can support large-scale networks with a

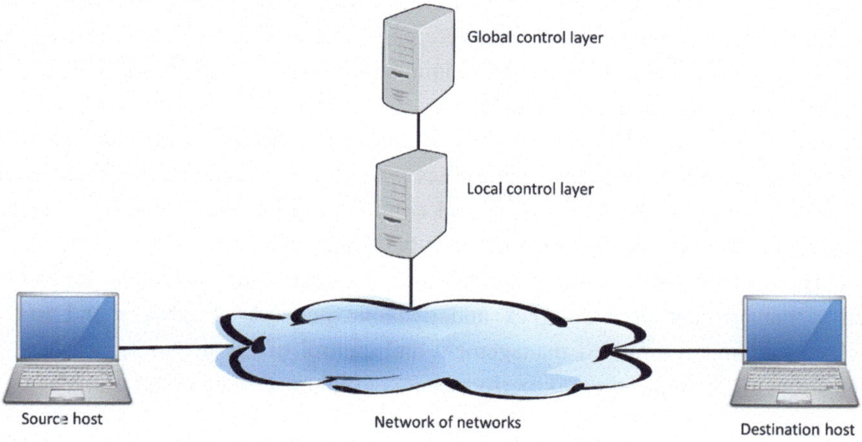

Fig. 2 Hierarchical controllers' architecture with two levels hierarchy

high degree of automation and flexibility. It also allows for faster response times and better fault isolation, as issues can be localized to specific regions or devices [21, 22].

Therefore, the hierarchical control layer SDN architecture is a network architecture that divides the network into multiple levels of SDN controllers, each responsible for managing a different aspect of the network. This architecture provides several benefits, including scalability, fault tolerance, and efficient resource utilization.

D. Infrastructure Layer

The infrastructure layer includes the physical devices and sensors that make up the SAG-IoT network. This layer includes a wide range of devices, including satellites, drones, ground sensors, and other sensor devices. Each device in the infrastructure layer is equipped with an SDN interface that allows it to communicate with the SDN controller via the southbound API and other devices on the network.

The SDN architecture for SAG-IoT is designed to be flexible and scalable, allowing organizations to add new devices and services to the network as needed. It is also designed to be secure, with robust authentication and encryption mechanisms to protect the network from unauthorized access.

Moreover, the SDN provides a more dynamic and programmable network infrastructure that enables the efficient management of IoT devices. It allows for centralized control of the network, which makes it easier to manage and configure the devices. SDN controllers can collect and analyze data from IoT devices to dynamically adjust network resources [23], including bandwidth and routing, to optimize the performance of the network.

In addition, the benefits of using SDN in IoT infrastructure include improved network scalability, reliability, and security. By leveraging SDN technologies, IoT devices can be managed more efficiently, and network administrators can automate many of the tasks associated with managing large-scale IoT.

networks. Similarly, SDN can be used to enable advanced IoT applications such as intelligent transportation systems, smart cities, and industrial automation. These applications require real-time data processing and communication, which can be achieved by using SDN technologies to optimize network performance.

Finally, the SDN architecture for SAG-IoT is designed to enable the efficient and effective management of a complex and dynamic network of devices and IoT sensors. By providing centralized management and control of the network, SDN can help organizations to optimize network performance and improve operational efficiency.

E. **Northbound and Southbound Interfaces**

The northbound interface refers to the interface between the SDN controller and the network applications or services. It enables communication between the network applications and the SDN controller, allowing the applications to request network services and policies, and receive information about the state of the network. The northbound interface is typically based on open APIs, such as REST APIs, which provide a standardized way for applications to interact with the SDN controller.

On the other hand, the southbound interface refers to the interface between the SDN controller and the network infrastructure devices, such as switches and routers. It enables the SDN controller to control the forwarding behavior of the network devices and manage the network topology. The southbound interface is typically based on protocols such as OpenFlow, which provide a standardized way for the SDN controller to communicate with the network devices. The SDN controller updates the rules known as flow rules in the devices of the data plane through this interface. Hence, updated or modified rules are communicated through OpenFlow or other protocols acting as a northbound interface.

Consequently, the northbound interface enables communication between the SDN controller and network applications, while the southbound interface enables communication between the SDN controller and network infrastructure devices. Both interfaces play a critical role in enabling the flexibility, programmability, and agility of software-defined networking.

F. **Using the Power of AI**

AI can play a critical role in software-defined networking (SDN) for Space-Air Ground IoT by providing intelligent automation, predictive analytics, and enhanced security. One way that AI can help in SDN for SAG-IoT is through intelligent automation. With AI, networks can be automatically configured and optimized in real-time based on changing conditions and requirements. This can improve network performance and reduce the need for manual intervention, which is essential in a dynamic environment like Space-Air Ground IoT.

Moreover, the AI can also help in SDN for SAG-IoT by providing predictive analytics. By analyzing data collected from the network, AI algorithms can identify patterns and trends that can help to predict future events or issues. For example, AI can be used to predict potential equipment failures or security breaches, allowing organizations to take proactive measures to prevent them from happening. Another way that AI can help in SDN for Space-Air Ground IoT is by enhancing security. With AI, networks can be monitored in real-time for potential threats, and anomalous behavior can be identified and flagged for further investigation. This can help to detect and prevent cyber-attacks, which are a significant concern in Space-Air Ground IoT.

Similarly, AI can also help to improve network optimization and performance in SDN for SAG-IoT. By analyzing network traffic and other data, AI algorithms can identify areas where the network can be optimized for better performance and efficiency. This can help to reduce latency, improve throughput, and increase overall network performance. In short, AI can play a critical role in SDN for SAG-IoT by providing intelligent automation, predictive analytics, enhanced security, and improved network optimization and performance. By leveraging the power of AI, organizations can create more efficient, effective, and secure networks for a wide range of SAG-IoT applications.

G. **Reinforcement Learning**

Herein, we explain how reinforcement learning will help in SDSAG-IoT [24]. reinforcement learning is a type of machine learning that enables an agent to learn how to interact with an environment through trial and error. In the context of SDSAG-IoT, reinforcement learning can be used to optimize the performance of the network by enabling the network to learn and adapt to changing conditions in real-time.

One of the key challenges in SDSAG-IoT is the need for the network to be highly adaptive and responsive to changing conditions. This includes changes in the environment, changes in the network topology, and changes in traffic patterns. Reinforcement learning can help address this challenge by enabling the network to learn and adapt to these changes in real time. For example, reinforcement learning can be used to optimize the routing of traffic in the network by learning the best paths to take based on the current network conditions.

This can include factors such as the current traffic load, the available bandwidth, and the latency of the network. By continuously learning and adapting to these conditions, the network can optimize its performance and provide a better quality of service to the users. Moreover, Reinforcement learning can also be used to optimize the allocation of network resources, such as bandwidth and processing power. By learning the optimal allocation of resources based on the current network conditions, the network can optimize its performance and reduce congestion and delays. Hence, reinforcement learning can help in SDSAG-IoT by enabling the network to learn and adapt to changing conditions in real-time. By continuously optimizing its performance, the network can provide a better quality of service to the users and improve the overall efficiency of the network.

In contemporary wireless communication networks, despite advancements, the surge in user-equipment surpasses network development, posing resource scarcity and challenges in meeting escalating traffic demands. Space-Air-Ground Integrated Networks (SAGIN) emerged as a potential remedy, amalgamating major networks to offer vast resource pools, flexibility, and extensive coverage. Yet, challenges persist, notably in managing traffic flow, necessitating a robust offloading scheme for this heterogeneous and dynamic environment. Existing conventional methods fall short due to static network designs. Prior research has explored traffic offloading in SAGIN-like setups, like satellite-based traffic distribution and machine learning-based offloading, albeit facing limitations in adapting to the complexities of SAGIN networks. To tackle these hurdles, our novel proposal integrates delay-sensitive algorithms, neighborhood information discovery protocols, and a machine learning model within a reinforcement learning-based traffic offloading system. This adaptable solution aims to address the intricacies of dynamic networks, offering scalability and flexibility, a departure from existing niche-specific approaches. The Chapter progresses to detail the background, related challenges, and the proposed distributed reinforcement learning-based traffic offloading system utilizing Double Deep Q-Learning in subsequent sections for a comprehensive understanding and performance evaluation [25].

The escalating demands for high-quality wireless services have strained traditional cellular networks, prompting the exploration of UAV-assisted communications. UAVs serving as aerial base stations offer swift wireless service recovery post-disasters, alleviate network congestion in hotspots, and facilitate massive connectivity in IoT networks. Their agility, cost-effectiveness, line-of-sight propagation, and mobility are key advantages, enabling quick deployment, affordable infrastructure, reduced signal interference, and flexible on-demand services. Notably, major telecom players like AT&T, Huawei, and China Mobile have delved into UAV-based LTE services, showcasing their potential at events like the Mobile World Congress. However, as the demand for UAVs in wireless networks grows exponentially, the challenges in designing efficient UAV-assisted networks become increasingly daunting [26].

The utilization of unmanned aerial vehicles (UAVs) for real-time sensing in cellular networks presents challenges in timely data transmission to the base station (BS). To address this, cellular network-controlled UAV transmission is proposed as a solution, ensuring uplink Quality of Service (QoS). However, determining UAV trajectories poses a significant challenge in maintaining both sensing accuracy and uplink transmission quality. This becomes complex in decentralized environments where UAVs from various entities compete for scarce spectrum resources. To resolve these issues, this Chapter introduces a sense-and-send protocol for coordinating real-time sensing UAVs, employing nested Markov chains to assess successful data transmission probability. It formulates a decentralized UAV trajectory design problem and proposes an enhanced multi-UAV Q-learning algorithm, exhibiting faster convergence and higher rewards compared to single-agent and opponent modeling Q-learning algorithms in simulation experiments [27].

This work addresses the need for autonomous UAV navigation in vast, intricate environments like urban areas through a deep reinforcement learning (DRL)-based technique, essential for applications like emergency aid, goods delivery, and surveillance. Various methods, from non-learning-based to machine learning approaches, have tackled UAV navigation problems. However, in complex, dynamic spaces, non-learning methods relying on explicit path planning or SLAM may falter. Some works resort to machine learning, like reinforcement learning (RL), yet face challenges in adapting to large-scale complex environments with limited sensors and continuous action spaces. To confront these issues, a partially observable Markov decision process (POMDP) is modeled for UAV navigation, demanding a model-free, continuous action space approach. A novel online DRL algorithm, Fast-RDPG, is proposed to efficiently tackle POMDPs, revolutionizing the way UAVs navigate in large-scale complex settings, evidenced by a stochastic large-scale environment simulation [28].

This survey delves into the transformative impact of Unmanned Aerial Vehicles (UAVs), commonly known as drones, across various domains, reflecting their escalating role in modern applications. Expanding from aerial photography to precision agriculture, environmental monitoring, and rescue operations, UAVs have gained significant momentum. With an estimated growth of the global UAV market from approximately U.S.$30.6 billion in 2022 to an anticipated U.S.$55.8 billion by 2030, at a Compound Annual Growth Rate (CAGR) of 7.8%, their versatility and market expansion remain undeniable. However, the limitations of single UAV-based platforms have led to the exploration of multi-UAV frameworks, propelling the interest in Multi-UAV Wireless Networks (MUWNs). These networks, enhancing coverage and service resources, introduce coordinated efforts among multiple UAVs, promising advancements in IoT, cellular communications, and edge computing. Addressing the challenges in orchestration, trajectory planning, and resource management, this survey advocates for the adoption of Reinforcement Learning (RL) and Deep Reinforcement Learning (DRL) to render MUWNs adaptable, efficient, and intelligent. The survey comprehensively reviews the applications, challenges, and future directions of RL-enabled MUWNs, categorizing cutting-edge applications into six vital areas and emphasizing the pressing need for advanced algorithms to navigate the complexities of these networks [29].

4 Challenges in Software-Defined Space-Air-Ground Iot

There are several challenges associated with implementing software-defined networking (SDN) in Space-Air Ground IoT.

- One of the main challenges is the heterogeneity of the devices and sensors that make up the network [30]. In SAG-IoT, there may be a wide range of devices and sensors with varying capabilities and requirements. This can make it difficult to create a unified SDN architecture that works for all devices on the network. To overcome this challenge, organizations may need to invest in specialized hardware and software that is specifically designed for SAG-IoT [31].
- Another challenge is the complexity of the network topology. In SAG-IoT, the network topology can be highly dynamic, with devices moving in and out of range, and changing their position in relation to other devices. This can make it difficult to create a stable and predictable network topology that can be managed using SDN. To address this challenge, organizations may need to develop new algorithms and protocols that can adapt to the changing network topology in real-time.
- Security is also a major challenge in Space-Air Ground IoT. With so many devices and sensors on the network, it can be difficult to ensure that all devices are secure and protected from cyber threats. SDN can help to improve security by enabling centralized management of network policies and configurations, but it is still important to implement robust security measures at the device level to prevent unauthorized access.
- Finally, the deployment of SDN in SAG-IoT may require significant investment in terms of time, resources, and expertise. Organizations may need to hire specialized personnel with expertise in both SDN and IoT and invest in specialized hardware and software to support the network. Besides the SDN can bring significant benefits to SAG-IoT, there are also significant challenges associated with its deployment. Organizations must be prepared to invest the necessary resources and expertise to overcome these challenges and achieve the full potential of SDN in SAG-IoT.

In today's mobile-centric era, the burgeoning demand for Quality of Service (QoS) and the advent of the fifth-generation (5G) wireless communication system are reshaping the landscape of connectivity. Communication networks extend beyond densely populated urban hubs, reaching remote and sparsely inhabited regions like oceans, forests, and farmland. The Internet of Things (IoT) and the proliferation of autonomous driving technologies further propel the need for seamless and independent communication across various terrains. However, the conventional placement of base stations by mobile network operators (MNOs) primarily caters to high-density areas, leaving vast coverage gaps of terrestrial broadband wireless signals. Efforts to bridge this gap have introduced concepts like the "space–air–ground-integrated networks" (SAGINs), envisioning a heterogeneous network comprising sky access platforms (SAPs) above terrestrial networks. While studies have explored aerial platforms' theoretical benefits in network coverage enhancement, they often overlooked the integration of civil aircraft (CAs) within SAGINs. Challenges loom large for implementing software-defined networking (SDN) in Space-Air Ground IoT, including device heterogeneity, dynamic network topology, security concerns, and substantial investments required in terms of time, resources, and expertise [32].

In recent years, the proliferation of low-cost smart devices and the exponential growth of IoT-connected devices have paved the way for transformative applications across various domains. Predictions suggest a substantial surge in connected devices and data volumes, indicating the need for communication networks that can meet diverse QoS requirements in urban and remote areas. While existing terrestrial networks like Wi-Fi, LTE, and 5G face limitations in providing reliable and cost-effective services, especially in extreme areas, the emergence of Space-Air-Ground Integrated Networks (SAGINs) offers a promising solution. SAGINs, comprising communication satellites, aerial devices such as balloons and UAVs, and terrestrial networks, promise extensive coverage and adaptable access for remote regions and on-demand applications. However, SAG-IoT's success is impeded by security, reliability, and privacy concerns arising from dynamic network topologies, multi-segment heterogeneity, and decentralized architectures. The integration of blockchain with SAG-IoT shows promise in mitigating these challenges by offering decentralized, transparent, and tamper-resistant security measures. Yet, the wide deployment of blockchain in SAG-IoT systems poses efficiency and scalability challenges, necessitating tailored solutions to address the complexity of multi-dimensional SAG-IoT networks. This survey delves into the security concerns in SAG-IoT, explores blockchain-based solutions, identifies potential applications, and outlines open research issues, distinguishing itself from previous surveys by emphasizing tailored blockchain solutions for securing SAG-IoT across space, air, and ground layers [33].

5 Conclusion and Future Directions

In this Chapter, we illustrate that the AI will play a crucial role in the success of SDN for SAG-IoT. AI can help to improve network automation, predictive analytics, security, and performance, making it an essential technology for the future of SAG-IoT. One of the significant benefits of AI in SDN for SAG-IoT is that it can provide intelligent automation. With AI, networks can be automatically configured and optimized in real-time based on changing conditions and requirements. This can help to improve network performance and reduce the need for manual intervention, which is critical in a dynamic environment like SAG-IoT. Another significant advantage of AI in SDN for Space-Air Ground IoT is its ability to provide predictive analytics. AI algorithms can analyze data collected from the network to identify patterns and trends that can help predict future events or issues.

For example, AI can be used to predict potential equipment failures or security breaches, allowing organizations to take proactive measures to prevent them from happening.

Despite the significant benefits, AI also poses some challenges for SDN in SAG-IoT. One of the significant challenges is the need for massive amounts of data to train AI algorithms effectively. This requires a robust data collection infrastructure, which can be challenging to implement in remote and harsh environments like SAG-IoT. Another significant challenge is the need for robust and secure data processing and storage. With AI, data privacy and security are critical, and organizations must ensure that data is adequately protected from cyber threats and unauthorized access.

Future research directions for AI in SDN for SAG-IoT could include developing more sophisticated algorithms for network optimization, security, and analytics. There is also a need for research on AI-powered fault tolerance and self-healing mechanisms to improve network reliability and reduce downtime. Moreover, AI has the potential to transform SDN in SAG-IoT, providing intelligent automation, predictive analytics, enhanced security, and improved network performance.

However, organizations must address the challenges and continue to invest in research and development to fully realize the potential of AI in this space.

Acknowledgements This work was supported by the MSIT (Ministry of Science and ICT), Korea, under the ITRC (Information Technology Research Center) support program (IITP-2022-2018-0-01431) supervised by the IITP (Institute for Information & Communications Technology Planning & Evaluation).

References

1. Weiwei Jiang, “Software defined satellite networks: A survey”, Digital Communications and Networks, 2023, ISSN 2352–8648, https://doi.org/10.1016/j.dcan.2023.01.016.
2. H. Babbar, S. Rani, S. Garg, G. Kaddoum, M. J. Piran and M. S. Hossain, "A Secure Multilayer Architecture for Software-Defined Space Information Networks," in IEEE Consumer Electronics Magazine, vol. 12, no. 2, pp. 64–72, 1 March 2023, https://doi.org/10.1109/MCE.2021.3139169.
3. N. Zhang, S. Zhang, P. Yang, O. Alhussein, W. Zhuang and X. S. Shen, "Software Defined Space-Air-Ground Integrated Vehicular Networks: Challenges and Solutions," in IEEE Communications Magazine, vol. 55, no. 7, pp. 101–109, July 2017 https://doi.org/10.1109/MCOM.2017.1601156.
4. Chi, Hao Ran, Chung Kit Wu, Nen-Fu Huang, Kim Fung Tsang, and Ayman Radwan. "A Survey of Network Automation for Industrial Internet-of-Things Towards Industry 5.0." IEEE Transactions on Industrial Informatics (2022).
5. L. Zhang, F. Liu and J. Sun, "Sofware-Defined Space-Ground-Vehicle Integrated Network Architecture for Unconscious Payment," in IEEE Access, vol. 10, pp. 128122–128131, 2022https://doi.org/10.1109/ACCESS.2022.3227396.

6. Nan CHENG, Jingchao HE, Zhisheng YIN, Conghao ZHOU, Huaqing WU, Feng LYU, Haibo ZHOU, Xuemin SHEN, "6G service-oriented space-air-ground integrated network: A survey", Chinese Journal of Aeronautics, Volume 35, Issue 9, 2022, Pages 1–18, ISSN 1000-9361, https://doi.org/10.1016/j.cja.2021.12.013.
7. D. Zhou, M. Sheng, J. Li and Z. Han, "Aerospace Integrated Networks Innovation for Empowering 6G: A Survey and Future Challenges," in IEEE Communications Surveys & Tutorials, https://doi.org/10.1109/COMST.2023.3245614.
8. J. Li, W. Shi, H. Wu, S. Zhang and X. Shen, "Cost-Aware Dynamic SFCMapping and Scheduling in SDN/NFV-Enabled Space–Air–Ground-Integrated Networks for Internet of Vehicles," in IEEE Internet of Things Journal, vol. 9, no. 8, pp. 5824–5838, 15 April15, 2022, https://doi.org/10.1109/JIOT.2021.3058250.
9. Xiaoyi Tao, Kaoru Ota, Mianxiong Dong, Heng Qi, Keqiu Li. "Congestion-Aware Scheduling for Software-Defined SAG Networks", IEEE Transactions on Network Science and Engineering, 2021.
10. S. Xu, X. -W. Wang and M. Huang, "Software-Defined Next-Generation Satellite Networks: Architecture, Challenges, and Solutions," in IEEE Access, vol. 6, pp. 4027–4041, 2018, https://doi.org/10.1109/ACCESS.2018.2793237.
11. Cheng, N., Jingchao, H. E., Zhisheng, Y. I. N., Conghao, Z. H. O. U., Huaqing, W. U., Feng, L. Y. U., ... & Xuemin, S. H. E. N. (2022). 6G service-oriented space-air-ground integrated network: A survey. Chinese Journal of Aeronautics, 35(9), 1–18.
12. Guo, C., Gong, C., Xu, H., Zhang, L., & Han, Z. (2022). A Dynamic Handover Software-Defined Transmission Control Scheme in Space-Air-Ground Integrated Networks. IEEE Transactions on Wireless Communications, 21(8), 6110–6124.
13. Wang, J., & Liu, J. (2021). Location Hijacking Attack in Software-Defined Space–Air–Ground-Integrated Vehicular Network. IEEE Internet of Things Journal, 9(8), 5971–5981.
14. Liu, Y., Jiang, L., Qi, Q., & Xie, S. (2022). Energy-Efficient Space–Air–Ground Integrated Edge Computing for Internet of Remote Things: A Federated DRL Approach. IEEE Internet of Things Journal, 10(6), 4845–4856.
15. Zhang, P., Chen, N., Shen, S., Yu, S., Kumar, N., & Hsu, C. H. (2023). AI-Enabled Space-Air-Ground Integrated Networks: Management and Optimization. IEEE Network.
16. Zhang, P., Li, Y., Kumar, N., Chen, N., Hsu, C. H., & Barnawi, A. (2022). Distributed deep reinforcement learning assisted resource allocation algorithm for space-air-ground integrated networks. IEEE Transactions on Network and Service Management.
17. Wu, H., Chen, J., Zhou, C., Shi, W., Cheng, N., Xu, W., ... & Shen, X. S. (2020). Resource management in space-air-ground integrated vehicular networks: SDN control and AI algorithm design. IEEE Wireless Communications, 27(6), 52–60.
18. Ali, J., Shan, G., Gul, N., & Roh, B. H. (2023). An Intelligent Blockchain-based Secure Link Failure Recovery Framework for Software-defined Internet-of-Things. *Journal of Grid Computing*, 21(4), 57.
19. Ali, Jehad, Rutvij H. Jhaveri, Mohannad Alswailim, and Byeong-hee Roh. "ESCALB: An effective slave controller allocation-based load balancing scheme for multi-domain SDN-enabled-IoT networks." *Journal of King Saud University-Computer and Information Sciences* 35, no. 6 (2023): 101566.
20. Ali, J. and Roh, B.H., 2022. An Effective Approach for Controller Placement in Software-Defined Internet-of-Things (SD-IoT). *Sensors*, *22*(8), p.2992.
21. C. Chen, Z. Liao, Y. Ju, C. He, K. Yu and S. Wan, "Hierarchical Domain-Based Multicontroller Deployment Strategy in SDN-Enabled Space–Air–Ground Integrated Network," in IEEE Transactions on Aerospace and Electronic Systems, vol. 58, no. 6, pp. 4864-4879, Dec. 2022, https://doi.org/10.1109/TAES.2022.3199191.

22. Sun, J., Liu, F., Li, Y., Zhang, L. and Shi, D., 2021. A Software-Defined Architecture for Integrating Heterogeneous Space and Ground Networks. Frontiers in Communications and Networks, 2, p.717476.
23. H. Wu et al., "Resource Management in Space-Air-Ground Integrated Vehicular Networks: SDN Control and AI Algorithm Design," in IEEE Wireless Communications, vol. 27, no. 6, pp. 52–60, December 2020, https://doi.org/10.1109/MWC.001.2000130.
24. Jiadai Wang, Jiajia Liu. "Location Hijacking Attack in Software-Defined Space–Air–Ground-Integrated Vehicular Network", IEEE Internet of Things Journal, 2022.
25. Tang, F., Hofner, H., Kato, N., Kaneko, K., Yamashita, Y., & Hangai, M. (2021). A deep reinforcement learning-based dynamic traffic offloading in space-air-ground integrated networks (SAGIN). IEEE Journal on Selected Areas in Communications, 40(1), 276–289.
26. Liu, X., Liu, Y., & Chen, Y. (2019). Reinforcement learning in multiple-UAV networks: Deployment and movement design. IEEE Transactions on Vehicular Technology, 68(8), 8036–8049.
27. Hu, J., Zhang, H., & Song, L. (2018). Reinforcement learning for decentralized trajectory design in cellular UAV networks with sense-and-send protocol. IEEE Internet of Things Journal, 6(4), 6177–6189.
28. Wang, C., Wang, J., Shen, Y., & Zhang, X. (2019). Autonomous navigation of UAVs in large-scale complex environments: A deep reinforcement learning approach. IEEE Transactions on Vehicular Technology, 68(3), 2124–2136.
29. Bai, Y., Zhao, H., Zhang, X., Chang, Z., Jäntti, R., & Yang, K. (2023). Towards Autonomous Multi-UAV Wireless Network: A Survey of Reinforcement Learning-Based Approaches. IEEE Communications Surveys & Tutorials.
30. Ali, J., & Roh, B. H. (2020). An effective hierarchical control plane for software-defined networks leveraging TOPSIS for end-to-end QoS class-mapping. *Ieee Access*, *8*, 88990–89006.
31. Weiwei Jiang. "Graph-based deep learning for communication networks: A survey", Computer Communications, 2022.
32. Niu, Z., Shen, X. S., Zhang, Q., & Tang, Y. (2020). Space-air-ground integrated vehicular network for connected and automated vehicles: Challenges and solutions. Intelligent and Converged Networks, 1(2), 142–169.
33. Wang, Y., Su, Z., Ni, J., Zhang, N., & Shen, X. (2021). Blockchain-empowered space-air-ground integrated networks: Opportunities, challenges, and solutions. IEEE Communications Surveys & Tutorials, 24(1), 160–209.

NOMA-Based Access and Edge Computing in mmWave Enabled Aerial and Ground Integrated Networks

Amara Umar and Syed Ali Hassan

Abstract

In contemporary times, there has been a notable surge in the exploration of pervasive coverage, high-speed data connectivity, and edge computing as pivotal components of prospective sixth-generation (6G) wireless networks. These services are regarded as key components vital for the expected progression in 6G technology. However, the effective realization of these services is heavily contingent upon the existence of resilient network coverage and telecommunication infrastructure. Regrettably, in remote and secluded areas, this infrastructure is frequently absent, presenting formidable obstacles in the attainment of seamless connectivity, extensive coverage, and optimal computation offloading. Targeting the aforementioned horizon, we propose a millimeter wave (mmWave) enabled non-orthogonal multiple-access (NOMA) based access and edge computing network in which high-altitude platforms (HAPs) provide access and edge computing services to ground user terminals. The mmWave spectrum offers huge bandwidth and NOMA due to its inherent benefits enhances system effective throughput and spectral efficiency in comparison to conventional access schemes. We analyze the performance of the proposed network and aim to minimize the execution delay experienced by users in a NOMA cluster offloading their data to edge computing servers located at the HAPs. For minimizing the execution delay, we devise a dual-layer optimization scheme aiming to minimize the transmission and computation delay by optimizing the transmission power and computational resource allocation. The optimization is performed by using Lagrange multipliers method and then convergence is attained by implementing a

A. Umar (✉) · S. A. Hassan
School of Electrical Engineering and Computer Science (SEECS), National University of Sciences & Technology (NUST), 44000 Islamabad, Pakistan
e-mail: aumar.dphd19seecs@seecs.edu.pk

© The Author(s), under exclusive license to Springer Nature Switzerland AG 2025

M. A. Jamshed and A. Nauman (eds.), *Integrated Terrestrial and Non-Terrestrial Networks*, Synthesis Lectures on Communications,
https://doi.org/10.1007/978-3-031-63705-6_5

sub-gradient approach. Optimal power allocation enhances the data rates which subsequently reduces the transmission delay while the optimal cores assignment greatly reduces the computation delay. Simulation results demonstrate the efficacy of our proposed approach, showing a notable reduction in execution delay.

Keywords
Non-orthogonal multiple access (NOMA) • Sixth-generation (6G) • Millimeter wave (mmWave) • High altitude platforms (HAPs) • Edge computing

1 Introduction

The emergence of fifth-generation (5G) systems has exhibited outstanding performance in terms of high data rates and extensive connectivity. This surge foresees a prominent increase in network applications and active devices, accompanied by a diverse array of services such as high-speed video streaming, augmented reality (AR), virtual reality (VR), and real-time interactive gaming. These services while demanding high spectral efficiency, high energy efficiency, low latency, and enhanced fairness present challenges due to their computational complexity and energy requirements. Their execution on mobile devices with constrained battery capacity and computational capabilities is arduous. Consequently, the cellular networks are undergoing continual evolution to meet the demands for capacity and computation, driven by the exponential growth in connected devices, demand for high data rates and computationally complex services.

In this context, devices with latency constraint and computationally complex applications can benefit from edge computing. Edge computing involves offloading computationally intensive data to powerful servers located at network edges thus reducing latency and enhancing computational efficiency. While the terrestrial edge computing framework generally offers effective quality-of-experience (QoE), it heavily depends on the telecommunication framework located in vicinity. This becomes challenging in far-flung areas with intricate and varied topography like deserts and mountains where maintaining and ensuring connectivity is difficult. Additionally, in such regions, network coverage is quite restricted and susceptible to disruptions caused by natural calamities, leading to unreliable network connectivity. Thus, the existing communication systems alone may not suffice to meet the rigorous requirements pertaining to network coverage, capacity, and computation. Consequently, both the industry and scholarly fraternity have turned their attention toward several key areas. Firstly, exploration of millimeter wave (mmWave) and Terahertz (THz) band is being conducted to unlock promising data rates [1]. Secondly, there is a focus on integrating spatial, aerial and terrestrial networks to extend services to unserved areas [2]. Moreover, emphasis is placed on exploiting the integrated networks for edge computing [3–7], while addressing the involved latency constraint for devices having computationally complex applications. Additionally, the design of unsupervised intelligent networks is being pursued, thus leading a path for the development of sixth-generation (6G) wireless networks. These efforts collectively aim to address the growing demands and shape the future of wireless communication [8, 9].

The quest for achieving multi-gigabit data rates has prompted research in mmWave frequency bands spanning from 10 to 300 gigahertz (GHz). Despite the hurdles posed by substantial path loss in mmWave communication, advancements in antenna design have led to the creation of compact, high-gain antennas capable of effectively surmounting these challenges. These advancements present an opportunity to harness the extensive bandwidth available in these underutilized frequency ranges while adhering to antenna power and size constraints. Recognizing the potential of spatial, aerial and ground integrated networks operating at mmWave frequencies, the 3rd Generation Partnership Project (3GPP) has outlined a comprehensive framework encompassing channel models, network parameters, and deployment scenarios for spatial and aerial communication systems [10].

High altitude platforms (HAPs), unmanned aerial vehicles (UAVs) or satellites act as network edges in aerial or orbital edge computing [11–13]. HAPs are stratospheric stations strategically positioned at a designated altitude, and maintaining a fixed position relative to the Earth's surface. These platforms are equipped with broad-spectrum capabilities, showcasing substantial potential in expanding connectivity and providing aerial edge computing services to challenging or under-served regions. Future applications of wireless networks are anticipated to involve sophisticated computations, intricate transmissions, and extensive data processing. Consequently, the aerial and ground integrated networks demand meticulous low-latency task processing schemes coupled with effective resource management to ascertain optimal functionality.

Non-orthogonal multiple-access (NOMA) stands as a promising paradigm within the realm of 5G and future cellular networks [14, 15]. This innovative approach facilitates concurrent service delivery to multiple users, utilizing the same radio resources while minimizing inter-user interference. By super-positioning the message signals in power domain and leveraging disparities in channel gains, NOMA effectively enhances bandwidth for individual user while allowing the scheduling of more users than the available resources. This technology exhibits advantages such as high cell edge throughput, reduced transmission latency, enhanced spectral efficiency, and improved user fairness. The combination of NOMA and edge computing where the paired users apply NOMA principles to efficiently offload their data onto edge servers has become a point of research attention, driven by the unique advantages offered by both methodologies [16].

In this chapter, we put forward a NOMA-based access and edge computing scenario for a mmWave enabled aerial and ground integrated networks where HAPs providing access and edge computing services to ground users. We evaluate the downlink performance of the proposed network on the basis of realistic channel specifications provided by 3GPP [10] and propose an optimization framework for latency minimization in uplink communication. The mmWave frequency band offers huge bandwidth that results in much reduced transmission latency while NOMA further improves system performance by enhancing the available bandwidth for each user. Integration of aerial and terrestrial networks helps in extending edge computing services to unserved and hard to reach places. A dual-layer optimization scheme is devised to reduce the latency of the proposed system. The Lagrangian dual decomposition method is utilized to optimize the transmission power of each user and the number of cores assigned to an individual user in a NOMA cluster. The optimization problem is solved by following a two-step process. Firstly, the Karush-Kuhn-Tucker (KKT) conditions are applied to the Lagrangian function, facilitating the determination of the Lagrangian multipliers. Subsequently, the convergence is achieved using the sub-gradient method. Optimal resource allocation minimizes and equalizes the execution delay of the paired user in a NOMA cluster and hence reduces resource wastage. Finally, the results are compared with no optimization case to demonstrate the effectiveness of the proposed model.

1.1 Literature Review

Recent research explorations have highlighted the potential of mmWave enabled aerial and ground integrated networks to complement terrestrial communication systems. A notable instance of this concept is demonstrated in the study conducted by researchers in [2], where they investigated the functionality of integrated networks operating at mmWave frequencies and explored various implementation scenarios and enabling technologies. Similarly, the viability of mmWave-enabled integrated networks was validated by authors in [17]. This study involved satellites equipped with enhanced-gain directional antennas and employed beamforming techniques to ascertain feasibility. Furthermore, in [18], an assessment of different cooperative architectures was presented, featuring multiple configurations involving geostationary equatorial orbit (GEO)-satellites, low earth orbit (LEO)-satellites, HAPs, and earth stations. Moreover, a space-air-ground integrated network (SAGIN) based on orthogonal frequency division multiplexing (OFDM) was introduced in [19]. This model incorporated mmWave-enabled aerial and spatial entities to complement small cells operating in the Terahertz (THz) spectrum. A comprehensive analysis of system performance highlighted the advantages of transitioning users from terrestrial setups to aerial or spatial units.

In recent years, there has been significant research interest in edge computing in spatial or aerial networks due to their potential benefits in the development of 5G and 6G communication. There has been a considerable focus on satellite-aided orbital edge computing (OEC) networks. In this regard, a network model involving software-defined networking-enabled satellite-aided edge computing to boost network performance was proposed in [20]. The optimization problems aimed at minimizing network latency for multiple offloading scenarios were successfully solved. Results demonstrated that the proposed architecture reduced network latency by 10.43% in comparison to the traditional approach of using only terrestrial edge computing. Recognizing the potential of OEC, an OEC task allocation algorithm based on the greedy strategy in LEO satellite network was proposed in [21]. A computational cost based performance analysis was performed and the results demonstrated improvement in terms of average delay and energy consumption. In [22], a hybrid cloud and edge computing enabled LEO satellite network was presented by the authors. The study investigated the offloading decisions with the aim to reduce the power consumption of terrestrial users, while ensuring compliance with multiple constraints. The authors used the binary variables relaxation method for finding the optimal solution with minimal complexity. In this work [23], the authors proposed an offloading scheme for minimizing the delay in an OEC network while considering multiple resource based constraints. By utilizing Lyapunov optimization theory, the original problem was transformed into multiple deterministic one-slot problems and joint optimization was performed in an iterative manner. In [24], a multi-tier computation infrastructure integrating cloud and edge computing was presented. The authors introduced a reinforcement learning (RL)-based scheme to explore offloading policies. Wang et al. presented an integrated network that incorporated dual-edge computation capabilities to minimize energy consumption and delay for all users in [25]. Edge computing servers were deployed on both terrestrial base stations and LEO satellites, and the proposed computation offloading problem was solved using the Hungarian algorithm. The authors in [26] investigated a computation offloading strategy for the OEC network. They explored local and OEC to minimize the offloading cost for users using game theory. In [27], Qiu et al. introduced a software-defined OEC network for the dynamic management of caching and computing resources. They proposed a deep Q-learning algorithm to address the joint resource allocation optimization problem.

Furthermore, there has been a significant surge in research interest regarding UAV/HAP-aided aerial edge computing (AEC) in recent times. In [28], the authors developed a system assisted by a cognitive edge computing enabled UAV serving as a relay as well as a computing node. The NOMA scheme was considered to serve users and a Rosen's gradient projection algorithm was proposed to reduce the computation delay. Numerical results demonstrated the effectiveness of the proposed algorithm. The authors in [29] studied a case of AEC where ground vehicles offloaded resource-intensive tasks to UAVs, HAPs, or a combination of both. The authors defined an optimization problem and applied queuing theory for optimal offloading. In [30], researchers explored a UAV-based AEC system for minimizing the energy consumption of users by jointly optimizing multiple factors using the successive convex approximation (SCA) method. In [31], a study was conducted on a wireless-powered UAV-enabled AEC system, where the UAV offered both energy and edge computing services to the users. The problem of maximizing the computation rate under both the partial and binary computation offloading modes was addressed by employing alternating algorithms. Chen et al. [32] put forward a hybrid computing model that integrated a UAV swarm in edge computing to ensure high quality of service (QoS) for computation-intensive applications. The authors also examined several real-world applications including smart cities. The authors in [33] investigated a UAV-assisted edge computing architecture, where a UAV acted as a computing server or a relay for offloading tasks to the access point. The objective was to minimize the weighted sum energy consumption of the UAV and users while satisfying multiple constraints. The authors proposed an alternating optimization algorithm that utilized the Lagrange duality method and SCA method to jointly optimize multiple parameters.

The rest of the chapter is organized as follows. Section 2 presents the system layout and Sect. 3 discusses the results. Finally, the conclusion and future directions are provided in Sect. 4.

2 The System Layout

As shown in Fig. 1, we consider an aerial and ground integrated network in which user terminals in a cellular dead zone lacking cellular coverage are served by HAPs. The HAPs provide access as well as edge computing services to users. The HAPs are equipped with edge servers having extended capabilities and computationally complex tasks are offloaded by user terminals to HAPs. We characterize the channel between HAPs and user terminals as per realistic channel specification provided by 3GPP. NOMA is used as the access technique

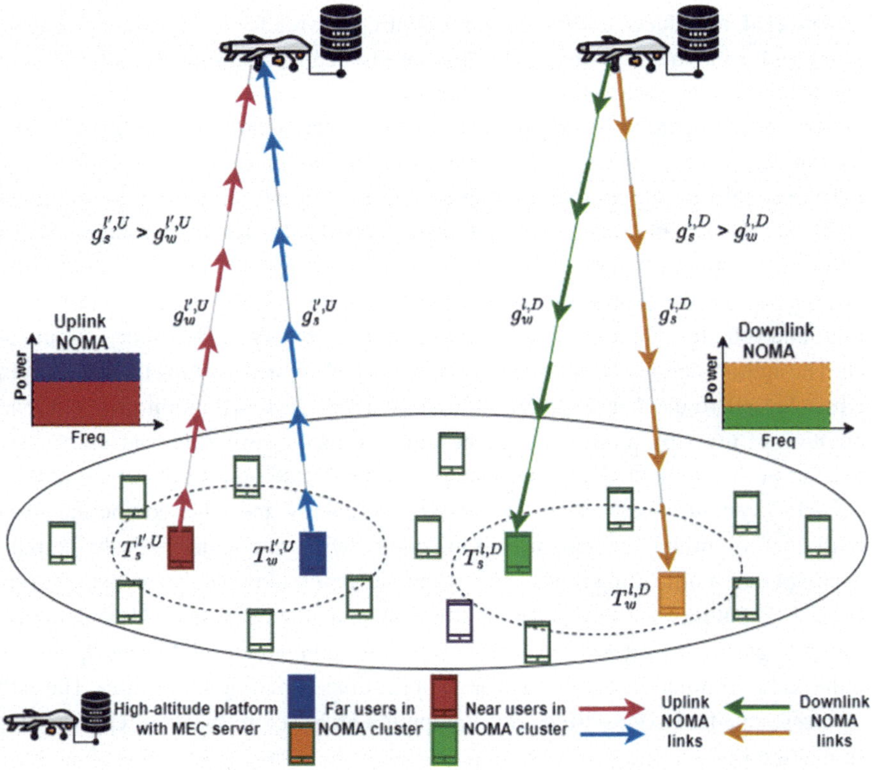

Fig. 1 The system model for NOMA-based aerial and ground integrated network

while mmWave range is targeted for both uplink and downlink communication. For efficient operation at mmWave range, it is assumed that HAPs support beamforming provided by large antenna arrays and the user terminals are compatible with this frequency range. We denote the set of user terminals in the network as $\mathcal{T} = \{1, 2, \ldots, T\}$ and the set of HAPs as $\mathcal{H} = \{1, 2, \ldots, H\}$, respectively. For both uplink and downlink NOMA communication, users are grouped into $\mathcal{L} = \{1, 2, \ldots, L\}$ clusters where $T > L$. Each cluster comprises of a strong (s) and a weak (w) users characterized on the basis of its channel conditions. For notation simplicity, we define $T_m^{l,k}$ as the mth user of cluster l, where $l \in L, m \in \{s, w\}$ and $k \in \{D, U\}$. Here, D/U denotes downlink/uplink communication. For both downlink and uplink communication, the total system bandwidth B is divided into C sub-channels, where $(C = B/L)$ and users in a cluster share the same sub-channel.

2.1 Pathloss Model

Path loss signifies the attenuation of a signal's strength as it traverses from a transmitting source to a receiving point. This fundamental principle in wireless communication is shaped by multiple variables, such as distance, frequency, environmental conditions, and obstacles. 3GPP specifies the path loss model for non-terrestrial networks operating at mmWave range [10]. According to these specifications, when a signal is transmitted from any non-terrestrial entity (satellite, HAP, UAV) towards a ground user, it undergoes several stages of attenuation represented as path loss. The path loss experienced by a signal propagating between a HAP and a user terminal is represented as

$$L_p[\text{dB}] = L_e + L_b + L_g + L_s, \tag{1}$$

where L_e is the building entry loss, which depends on the propagation mode and the elevation angle (α_e). The propagation mode can be line of sight (LoS) or non-line of sight (NLoS). The value of α_e also indicates the propagation mode with high value of α_e corresponding to LoS communication . The variable α_e depends on height of HAP (h_e) and the horizontal distance between the HAP and user terminal (d_h), as shown

$$\alpha_e = \arctan\left(\frac{h_e}{d_h}\right). \tag{2}$$

L_b is the basic path loss that caters for the free-space path loss (L_f), shadow fading (F_s) and clutter loss (L_c) as shown

$$L_b = L_f + F_s + L_c, \tag{3}$$

L_f depends on the carrier frequency (f_c) in GHz, and the distance d (also known as slant range) between a user terminal and a HAP present at height h_e as shown

$$L_f = 32.45 + 20\log(f_c) + 20\log(d). \tag{4}$$

The value of d is computed as

$$d = \sqrt{R_e^2 \sin^2(\alpha_e) + h_e^2 + 2h_e R_e} - R_e \sin(\alpha_e), \tag{5}$$

here R_e signifies Earth's radius. Similarly, F_s is a log-normal random variable with zero mean and standard deviation $\sigma_{\mathrm{F_s}}$. L_c is due to the attenuation caused by the surrounding obstructions and buildings, and its value varies with f_c and α_e. In the case of LoS propagation, and $\alpha_e = 90°$, L_c can be set to 0 dB.

Moreover, L_g is the attenuation due to gases, and it depends on zenith attenuation (A_z) and α_e as shown

$$L_g = \frac{A_z}{\sin(\alpha_e)}, \tag{6}$$

where A_z depends on f_c, and under normal atmospheric conditions its value is lesser than 10 dBs.

Finally, L_s represents tropospheric scintillation, which is observed at frequencies above 6 GHz. It occurs due to the changes in atmospheric refractive index, and its value increases with increasing f_c and reducing α_e.

2.2 Fading

Multipath fading is a prevalent occurrence in wireless communication systems. It is observed when transmitted signals from a sender arrive at the receiver by traversing numerous routes caused by reflection, diffraction, and scattering from buildings and obstructions in the surrounding. The outcome is fluctuations in signal amplitude and phase at the receiver, potentially causing signal deterioration and presenting performance issues. In our model, for the case of air-to-ground communication all the channels are assumed to follow Rician fading model [18], and its probability density function is given as

$$\begin{aligned} q_R(r) &= \frac{2r(R+1)}{\rho_r} \exp\left(-R - \frac{(R+1)r^2}{\rho_r}\right) \\ &\cdot I_o\left(2r\sqrt{\frac{R(R+1)}{\rho_r}}\right), r \geq 0, \end{aligned} \tag{7}$$

where ρ_r is the average envelop power, I_o is the zero order Bessel function, and R is the Rician factor, which is the ratio of power in specular and scatter components (Table 1).

2.3 Downlink NOMA

Considering the downlink communication, the channel coefficients for the strong and weak user of cluster l connected with HAP h over sub-channel c are $g_s^{l,D}$ and $g_w^{l,D}$, where $|g_m^{l,D}|^2$

Table 1 Notations and symbols

Notation	Definition	Notation	Definition
T	Total number of user terminals in the network	$B^{l,D}$	Bandwidth of sub-channel c allocated to cluster l in downlink
H	Total number of HAPs in the network	$B^{l',U}$	Bandwidth of sub-channel c' allocated to cluster l' in uplink
L	Total number of NOMA clusters in the network	$R_m^{l,D}$	Achievable data rate for m^{th} user of cluster l in downlink
$T_m^{l,D}$	m^{th} user of NOMA cluster l in downlink communication	$R_m^{l',U}$	Achievable data rate for m^{th} user of cluster l' in uplink
$T_m^{l',U}$	m^{th} user of NOMA cluster l' in uplink communication	$p_{h,m}^{l}[c]$	Power allocated to m^{th} user of cluster l over the sub-channel c by HAP h
h_e	Height of HAP	$p_m^{l'}[c']$	Transmission power allocated to m^{th} user of cluster l' over the sub-channel c'
d_h	Horizontal distance between HAP and user terminal	$d_m^{l'}$	Offloaded data size of mth user of cluster l'
d	Distance between HAP and user terminal (slant range)	$cx_m^{l'}$	Computational complexity of $d_m^{l'}$
f_c	Carrier frequency	$\eta_m^{l'}$	Computational resources allocated to m^{th} user of cluster l'
$g_m^{l,D}$	Channel coefficient for m^{th} user of cluster l in downlink	C_T	Number of CPU cores present at HAP
$g_m^{l',U}$	Channel coefficient for m^{th} user of cluster l' in uplink	f	Computational capability of each core
B	Total system bandwidth	C_t	Number of CPU cores allocated to a NOMA cluster

corresponds to Rician fading. The super-imposed signal transmitted by HAP h to cluster l over sub-channel c is

$$S_h^l[c] = \sqrt{p_{h,m}^l[c]} s_{h,m}^l[c], \tag{8}$$

where $p_{h,m}^l[c]$ is the power allocated to the mth user of cluster l over sub-channel c by HAP h and it is the sum of powers allocated to the strong and weak users, respectively. Downlink NOMA power allocation principle implies that the highest channel gain user gets the lowest power and vice versa. For the users in cluster l, as $|g_s^{l,D}|^2 > |g_w^{l,D}|^2$, hence $p_{h,s}^l < p_{h,w}^l$. Similarly in (8), $s_{h,m}^l[c]$ is the modulated signal for m^{th} user of cluster l transmitted over the sub-channel c by HAP h and it is considered to have zero mean is considered to have zero mean and unit variance, i.e., $\mathbb{E}\{|s_{h,m}^l[c]|^2\} = 1$.

The superimposed signal received by the strong and weak users of cluster l over sub-channel c is

$$I_m^l[c] = \frac{g_m^{l,D}}{\sqrt{L_p(d)}} S_h^l[c] + z^l[c], m \in \{s, w\}, \tag{9}$$

where $z^l[c] \sim \mathcal{CN}(0, \sigma^2)$ is the additive white Gaussian noise (AWGN) experienced by the users of cluster l over the sub-channel c, $L_p(d)$ is the path loss experienced at mmWave frequencies and $d \in \{d_{h,s}^l, d_{h,w}^l\}$, i.e., the distance between the strong or weak user and HAP h.

On the receiver side, the weak user $(T_w^{l,D})$ with channel gain $|g_w^{l,D}|^2$ directly decodes its signal by considering the strong user's $(T_s^{l,D})$ signal as interference. On the other hand, $T_s^{l,D}$ performs successive interference cancellation (SIC) to cancel out the interference caused by $T_w^{l,D}$ and decodes its signal without interference. The SINR for the corresponding strong and weak user of cluster l is

$$SINR_s^{l,D}[c] = \frac{p_{h,s}^l[c]|g_s^{l,D}|^2}{z^l[c]}, \tag{10}$$

$$SINR_w^{l,D}[c] = \frac{p_{h,w}^l[c]|g_w^{l,D}|^2}{p_{h,s}^l[c]|g_s^{l,D}|^2 + z^l[c]}. \tag{11}$$

The achievable data rate for the corresponding strong and weak users of cluster l is represented as

$$R_m^{l,D}[c] = B^{l,D} log_2(1 + SINR_m^{l,D}[c]), m \in \{s, w\}, \tag{12}$$

here $B^{l,D}$ corresponds to the bandwidth of sub-channel c allocated to cluster l. Consequently, the sum rate of cluster l is

$$R_{sum}^{l,D}[c] = \sum_{m \in \{s,w\}} R_m^{l,D}. \tag{13}$$

The total network sum rate in downlink is

$$R_{total}^{D} = \sum_{l=1}^{L} R_{sum}^{l,D}. \tag{14}$$

2.4 Uplink NOMA

In our model, HAPs provide edge computing services to ground user terminals. Users donot perform any local computations and offload their computationally extensive data processing tasks to HAPs via uplink NOMA communication. Similar to downlink NOMA, we consider a cluster l' comprising of 2 users, $T_s^{l',D}$ and $T_w^{l',D}$ in this study. Execution delay (D_e) is the delay incurred in sending and processing the computationally extensive data at the HAP. It comprises of the transmission delay (D_t) and the computation delay (D_c). For a NOMA cluster l' with 2 user, D_e for any m^{th} user is represented as

$$D_{e_m}^{l'} = D_{t_m}^{l'} + D_{c_m}^{l'}, m \in \{s, w\}, \tag{15}$$

where s and w are the respective strong and weak user. The value of D_{t_m} depends on the offloaded data size $(d_m^{l'})$ and achievable data rate $(R_m^{l',U})$. The value of D_{c_m} depends on $d_m^{l'}$, computational complexity of $d_m^{l'}$ i.e., $(cx_m^{l'})$, computation capability of each core (f) and the allocated computation resources $(\eta_m^{l'})$. Hence (15) can be rewritten as

$$D_{e_m}^{l'} = \frac{d_m^{l'}}{R_m^{l',U}} + \frac{d_m^{l'} cx_m^{l'}}{\eta_m^{l'} f}, m \in \{s, w\}. \tag{16}$$

Lets consider that the HAP is equipped with C_T number of cores. These cores are to be distributed among users for task processing. The cores allocated to a NOMA cluster l' is (C_t) where $C_t = C_T/L$. Value of C_t represents the number of cores shared by the strong $(T_s^{l',U})$ and weak $(T_w^{l',U})$ users of cluster l'. The computational complexity of offloaded data is contingent upon the data type, with multimedia data requiring more processing cycles compared to textual data. For a NOMA cluster, $d_m^{l'}$, $cx_m^{l'}$ and f are fixed, hence $D_{t_m}^{l'}$ depends on achievable data rate $R_m^{l',U}$ and consequently on allocated power $p_m^{l'}$, whereas $D_{c_m}^{l'}$ relies on the computational resources $\eta_m^{l'}$.

Considering the uplink communication, the channel coefficients for the strong and weak user of cluster l' over the sub-channel c' are $g_s^{l',U}$ and $g_w^{l',U}$, where $|g_m^{l',U}|^2$ corresponds to Rician fading. As the channel gain of strong user is higher in comparison to the weak one ($|g_s^{l',U}|^2 > |g_w^{l',U}|^2$), hence, as per the uplink NOMA principle, data of user $T_s^{l',U}$ will be decoded on priority at the HAP, the achievable data rate of which will be affected by interference from user $T_w^{l',U}$. However, $T_w^{l',U}$ performs SIC to cancel out the interference caused by $T_s^{l',U}$ and decodes its signal without interference. Hence the achievable data rate for $T_w^{l',U}$ will be effected by noise only. The achievable data rates rely on effective channel gains i.e., ($|g_s^{l',U}|^2$, $|g_w^{l',U}|^2$) and the assigned transmission powers i.e., ($p_s^{l'}[c']$, $p_w^{l'}[c']$) over the sub-channel c'. The SINR for the corresponding strong and weak user of cluster l' is

$$SINR_s^{l',U}[c'] = \frac{p_s^{l'}[c']|g_s^{l',U}|^2}{p_w^{l'}[c']|g_w^{l',U}|^2 + z^{l'}[c']}, \tag{17}$$

$$SINR_w^{l',U}[c'] = \frac{p_w^{l'}[c']|g_w^{l',U}|^2}{z^{l'}[c']}, \tag{18}$$

where $z^{l'}[c'] \sim \mathcal{CN}(0,\ \sigma^2)$ is the AWGN experienced by the users of cluster l' over the sub-channel c'. The achievable data rate for the corresponding strong and weak user of cluster l' is

$$R_m^{l',U}[c'] = B^{l',U}\log_2\left(1 + SINR_m^{l',U}[c']\right), m \in \{s, w\}, \tag{19}$$

here $B^{l',U}$ corresponds to the bandwidth of sub-channel c' allocated to cluster l'. Our objective is to minimize the execution delay of a NOMA cluster, consequently enhancing the resource allocation efficiency by reducing the delay difference among paired users. As illustrated in Fig. 2, a new execution process cannot commence until a user with a substantial load of computationally complex data finishes its ongoing transmission and computation process. It is important to note that the execution delay depends on the allocated power, $p_m^{l'}$, and the computational resources, $\eta_m^{l'}$. Therefore, through the optimization of these parameters, we can achieve an overall execution delay minimization.

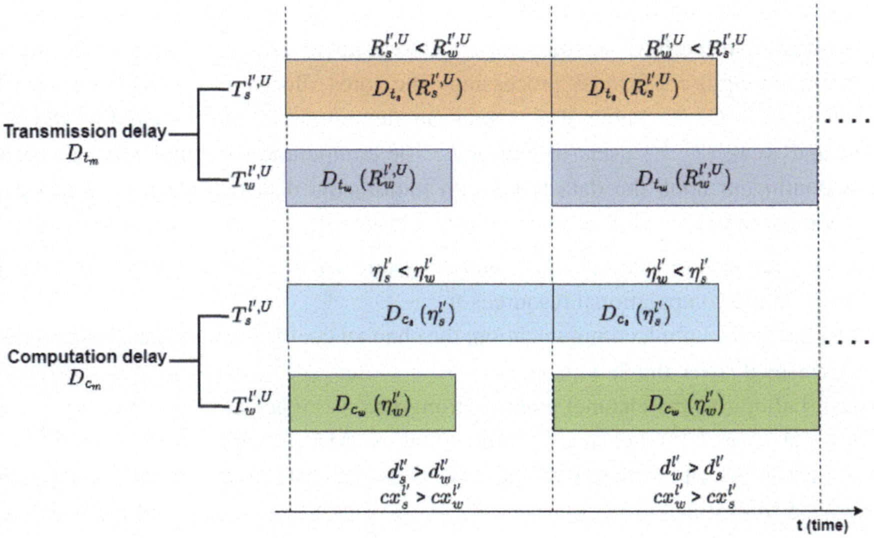

Fig. 2 An illustration of unequal execution delay and wasteful resources

2.4.1 Problem Formulation

The execution delay incured by a NOMA cluster is contingent upon the size of offloaded data ($d_s^{l'}, d_w^{l'}$) and its computational complexity ($cx_s^{l'}, cx_w^{l'}$). For the transmission delay that depends on the allocated power, consider a scenario that involves $d_s^{l'} > d_w^{l'}$ as shown in Fig. 2, if the power allocated to $T_s^{l',U} < T_w^{l',U}$, it leads to $R_s^{l',U} < R_w^{l',U}$ that eventually results in $D_{t_s} > D_{t_w}$. As shown, $T_w^{l',U}$ completes its transmission quite early but a new transmission cannot be initiated until $T_s^{l',U}$ also completes its transmission process. Similarly for the computation delay that depends on the number of allocated cores, consider a scenario in which $d_w^{l} > d_s^{l'}$ and $cx_w^{l'} > cx_s^{l'}$ as shown in Fig. 2, if the number of cores allocated to $T_s^{l',U} > T_w^{l',U}$, it results in $D_{c_w} > D_{c_s}$. As shown, $T_s^{l',U}$ gets done with its computation process quite early, however, a new computation process cannot be initiated until $T_w^{l',U}$ also completes its computation process. Since the users are paired, the allocated resources are released only when both the users in a cluster successfully conclude their execution involving both the transmission and computation process. This requires optimal resource allocation derived on the basis of size and computational complexity of offloaded data which eventually leads to the minimization as well as equalization of execution delay.

For minimizing the execution delay, it is essential to minimize D_{To}, which represents the maximum of $D_{e_m}^{l'}, m \in \{s, w\}$ for a NOMA cluster l', represented as

$$D_{To}(R_m^{l',U}, \eta_m^{l'}) = \max\left[D_{e_s}^{l'}(R_s^{l',U}, \eta_s^{l'}), D_{e_w}^{l'}(R_w^{l',U}, \eta_w^{l'})\right], \tag{20}$$

$$= \max\left[\left(\frac{d_s^{l'}}{R_s^{l',U}} + \frac{d_s^{l'} cx_s^{l'}}{\eta_s^{l'} f}\right), \left(\frac{d_w^{l'}}{R_w^{l',U}} + \frac{d_w^{l'} cx_w^{l'}}{\eta_w^{l'} f}\right)\right]. \tag{21}$$

Therefore, the objective function for addressing the delay minimization problem for a NOMA cluster is defined as follows

$$\begin{aligned}
&\text{minimize} && D_{To}(R_m^{l',U}, \eta_m^{l'}), && && \text{(22a)}\\
&\text{subject to} && C1 : p_m^{l'} \leq p_m^{l',\max} && m \in \{s, w\}, && \text{(22b)}\\
& && C2 : \eta_m^{l'} \leq C_t && m \in \{s, w\}, && \text{(22c)}\\
& && C3 : \eta_s^{l'} + \eta_w^{l'} \leq C_t, && && \text{(22d)}
\end{aligned}$$

here $C1$ constrains the power assigned to each user within their respective maximum limits. Additionally, $C2$ guarantees that the allocation of cores to a user does not surpass the total core count, C_t. Lastly, $C3$ dictates that the collective assignment of cores to both users must not exceed C_t.

We can solve the above optimization problem through the Lagrangian function which is given as

$$\mathcal{L}(p_m^{l'}, \eta_m^{l'}, \mu, \lambda, \theta) = D_{To}(R_m^{l',U}, \eta_m^{l'}) + \mu_m(p_m^{l',\max} - p_m^{l'}) + \lambda_m(C_t - \eta_m^{l'}) + \theta(Ct - \eta_s^{l'} - \eta_w^{l'}), \tag{23}$$

where $\mu_m = \{\mu_s, \mu_w\}$ represents the Lagrange multiplier linked with power allocation to users within the NOMA cluster. The parameters $\lambda_m = \{\lambda_s, \lambda_w\}$ and θ denote the Lagrange multipliers pertaining to cores allocation at the edge server. The Lagrangian function above is tackled through a two-step procedure. The initial step involves determining the optimal power that governs the data rates ($R_m^{l',U}$), and the optimal core count ($\eta_m^{l'}$) using the KKT conditions [34]. Subsequently, the Lagrange multipliers are found through the sub-gradient method. These KKT conditions represent the necessary first-order pre-requisites for identifying the optimal solution in a nonlinear, constrained optimization problem.

Next, the above Lagrangian function's partial derivatives with respect to $p_m^{l'}$ and $\eta_m^{l'}$ are equated to zero in order to derive the optimal power and core allocations for users in a NOMA cluster.

$$\left.\frac{\partial \mathcal{L}(p_m^{l'}, \eta_m^{l'}, \mu, \lambda, \theta)}{\partial p_m^{l'}}\right|_{p_m^{l'} = p_m^{l'*}} = 0, \tag{24}$$

and

$$\left.\frac{\partial \mathcal{L}(p_m^{l'}, \eta_m^{l'}, \mu, \lambda, \theta)}{\partial \eta_m^{l'}}\right|_{\eta_m^{l'} = \eta_m^{l'*}} = 0, \tag{25}$$

such that $p_m^{l'*}$ is the optimal transmit power and $\eta_m^{l'*}$ is the optimal number of cores for user m in a NOMA cluster. We obtain

$$\frac{\partial \mathcal{L}}{\partial p_s^{l'}} = \frac{(0.7)d_s^{l'}(p_s^{l'}|g_s^{l',U}|^2 + p_w^{l'}|g_w^{l',U}|^2 + z^{l'}[c'])}{B^{l',U}|g_s^{l',U}|^2} - \mu_s = 0, \tag{26}$$

$$\frac{\partial \mathcal{L}}{\partial p_w^{l'}} = \frac{(0.7)d_w^{l'}(p_w^{l'}|g_w^{l',U}|^2 + z^{l'}[c'])}{B^{l',U}|g_w^{l',U}|^2} - \mu_w = 0, \tag{27}$$

$$\frac{\partial \mathcal{L}}{\partial \eta_s^{l'}} = \theta - \lambda_s - \frac{d_s^{l'} c x_s^{l'}}{\eta_s^{l'2} f} = 0, \tag{28}$$

$$\frac{\partial \mathcal{L}}{\partial \eta_w^{l'}} = \theta - \lambda_w - \frac{d_w^{l'} c x_w^{l'}}{\eta_w^{l'2} f} = 0. \tag{29}$$

Accordingly, the optimal transmit power for users in a NOMA cluster is given as

$$p_s^{l'} = \left[\frac{\mu_s B^{l',U}|g_s^{l',U}|^2 - (0.7)(p_w^{l'}|g_w^{l',U}|^2 d_s^{l'} - z^{l'}[c'] d_s^{l'})}{(0.7)d_s^{l'}|g_s^{l',U}|^2}\right]^+, \tag{30}$$

$$p_w^{l'} = \left[\frac{\mu_w B^{l',U} |g_w^{l',U}|^2 - (0.7) z^{l'} [c'] d_w^{l'}}{(0.7) d_w^{l'} |g_w^{l',U}|^2} \right]^+. \tag{31}$$

Further, the optimal numbers of cores are given by

$$\eta_s^{l'} = \left\lfloor \left[\sqrt{\frac{d_s^{l'} c x_s^{l'}}{(\theta - \lambda_s) f}} \right]^+ \right\rfloor, \tag{32}$$

$$\eta_w^{l'} = \left\lfloor \left[\sqrt{\frac{d_w^{l'} c x_w^{l'}}{(\theta - \lambda_w) f}} \right]^+ \right\rfloor, \tag{33}$$

here, $[x]^+ = \max(0, x)$ and $\lfloor [x]^+ \rfloor$ represents floor operation as $\eta_m^{l'}$ is a positive integer. Moving forward, we compute the partial derivative of (23) with respect to the Lagrange multipliers linked to power and cores assignment, we obtain the following

$$\frac{\partial \mathcal{L}}{\partial \mu_m} = p_m^{l',\max} - p_m^{l'} = 0, \tag{34}$$

$$\frac{\partial \mathcal{L}}{\partial \lambda_m} = C_t - \eta_m^{l'} = 0, \tag{35}$$

$$\frac{\partial \mathcal{L}}{\partial \theta} = C_t - \eta_s^{l'} - \eta_w^{l'} = 0. \tag{36}$$

The Lagrangian multipliers μ, λ, and θ are derived utilizing the sub-gradient method and are presented as follows

$$\mu_m(i+1) = \left[\mu_m(i) + s_1 (p_m^{l',\max} - p_m^{l'}) \right]^+, \tag{37}$$

$$\lambda_m(i+1) = \left[\lambda_m(i) + s_2 (C_t - \eta_m^{l'}) \right]^+, \tag{38}$$

$$\theta(i+1) = \left[\theta(i) + s_3 (C_t - \eta_s^{l'} - \eta_w^{l'}) \right]^+, \tag{39}$$

where s_1, s_2, and s_3 denote the step sizes used for the iterative updates of the Lagrangian multipliers until convergence, while i signifies the iteration number. It is important to note that μ_m, λ_m and θ are always positive represented by $[.]^+$ function.

Algorithm 1 explains the proposed procedure for power and cores allocation to a NOMA cluster. The algorithm takes inputs that include initial values of Lagrangian multipliers $(\mu_m, \lambda_m, \theta)$, offloaded data specifications $(d_m^{l'}, cx_m^{l'})$, channel parameters $(|g_w^{l',U}|^2, B^{l',U})$ and core's computation capability (f). After a sequence of steps, the algorithm provides the optimal values of power and cores for the NOMA cluster. To initiate the process, the Lagrangian multipliers are set to their initial values, and step sizes are defined for iterative Lagrange multiplier computation. In this context, the variable i denotes the iteration number,

Algorithm 1 Power and Cores Allocation to NOMA Cluster

Input: $\mu_m, \lambda_m, \theta, d_m^{l'}, cx_m^{l'}, |g_w^{l',U}|^2, B^{l',U}, f, z^{l'}[c']$.
Output: $p_s^{l'}, p_w^{l'}, \eta_s^{l'}, \eta_w^{l'}$.

1: Set $\mu_m = \lambda_m = \theta = 0.01$, $s_1 = s_2 = s_3 = 0.01$, $i = 1$, $i_{\max} = 10^5$, $\epsilon = 0.01$.
2: **while** μ_m, λ_m, θ have not converged or $i \leq i_{\max}$ **do**
3: Calculate $p_w^{l'}$ from (31).
4: Using value of $p_w^{l'}$, calculate $p_s^{l'}$ from (30).
5: Calculate $\eta_s^{l'}$ and $\eta_w^{l'}$ from (32) and (33).
6: Update μ_m, λ_m, θ from (37, 38, 39).
7: **end while**

with $i_{\max}$ representing the maximum iterations. Step 2 specifies that the algorithm will continue to run until the Lagrangian multipliers have not yet converged or until the maximum iteration limit is reached. The convergence criterion is established as ϵ, indicating that the algorithm will persist until the difference between two consecutive multiplier values falls below ϵ. The optimal values of power and cores are determined upon convergence. With a fixed step size, our algorithm achieves convergence within a specified range of the optimal value [34].

3 Simulation Results and Analysis

In this section we evaluate the proposed model with reference to its downlink and uplink performance. The simulation setup involves $H = 2$ HAPs operating at an altitude of 20 Km above earth surface and $T = 200$ user terminals present randomly in 1000×1000 m^2 area. The simulation parameters are listed in Table 2.

The total network sum rate in downlink, R_{total}^{D} is shown in Fig. 3. The bias factor ϕ (dBs) acts as an additional gain to overcome the losses and enhance achievable rate. The mmWave band offers a huge bandwidth which results in attaining a sum rate in Gbps. It is observed that network sum rate in Gbps is attained without adding any extra gain ($\phi = 0$) when the HAPs are transmitting with their initially assigned EIRP value. The network sum rate shows an increasing trend with every increase in ϕ.

Figure 4 shows the execution delay experienced by the users in NOMA cluster l'. The offloaded data size ratio ($d_s^{l'}/d_w^{l'}$) and computational complexity ratio ($cx_s^{l'}/c_w^{l'}$) are varied to investigate their impact on the execution delay. Initially, $d_s^{l'} < d_w^{l'}$ as indicated by the ratio value lesser than 1 in Fig. 4 and $d_s^{l'} > d_w^{l'}$ after the ratio value surpasses 1. It is observed that execution delay is very high without any optimization. However, optimal power and

Table 2 Simulation parameters

Parameter	Value	Parameter	Value
Number of user terminal, T	200	Bandwidth, B	400 MHz
Number of HAPs, H	2	Fading, R	10
Downlink frequency, f_C	20 GHz	Noise figure	9 dB
Uplink frequency, f_C	30 GHz	CPU cores per cluster, C_t	20
HAP altitude, h	20 Km	Core computational capability, f	6 GHz
HAP EIRP	27.9 dB	Amount of offloaded bits, $d_m^{l'}$	[800–11,000] Kbits
User terminal transmission power, $p_m^{l'}$	[2, 4] Watts	Complexity of offloaded bits, $cx_m^{l'}$	[200–2500] Mcycles/Kbit

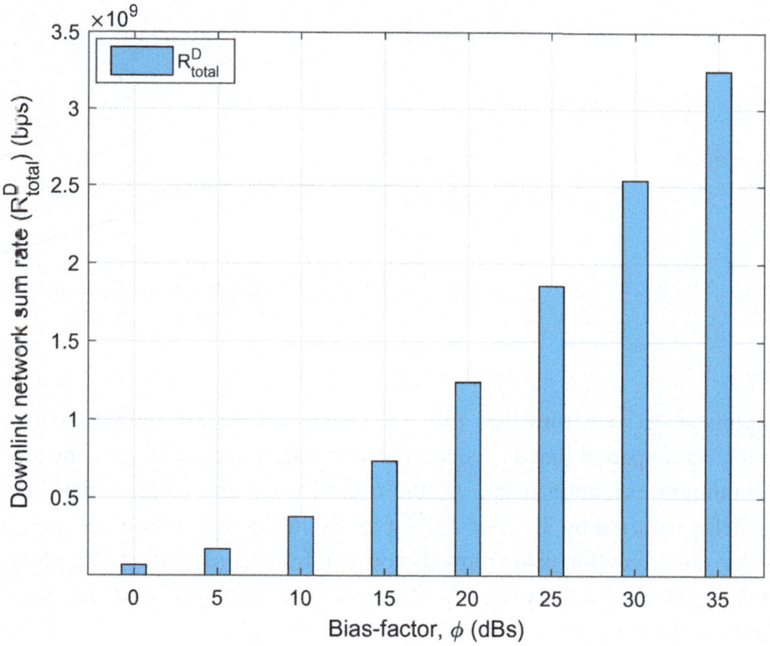

Fig. 3 Simulation and optimization results comparison

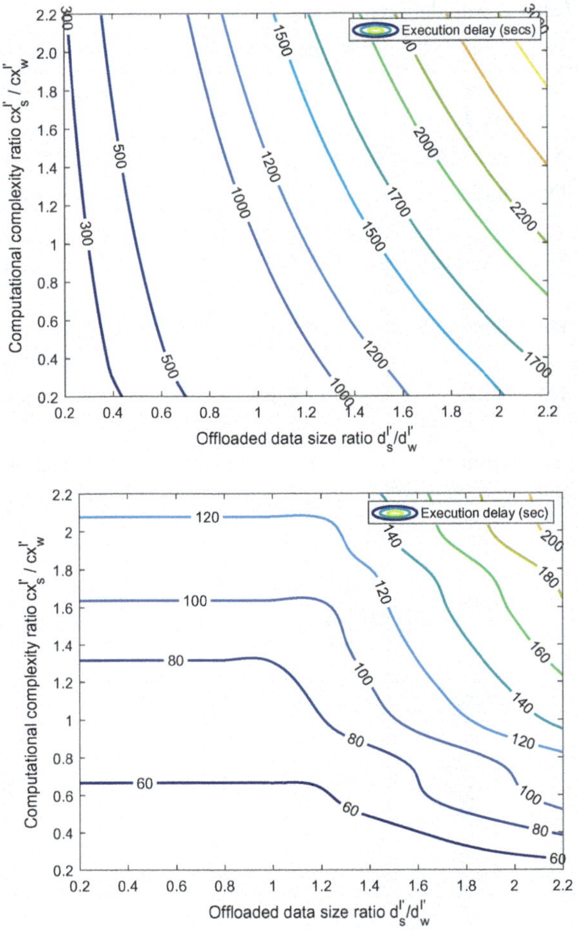

Fig. 4 Execution delay for no optimization and power and cores optimization

cores assignment derived from Eqs. (30)–(33) substantially reduces the execution delay. Additionally, it is apparent that a cluster attains optimal values of $d_s^{l'}/d_w^{l'}$ and $cx_s^{l'}/c_w^{l'}$, that result in minimum execution delay. As the overall execution delay is dependent on the maximum delay incurred by $T_s^{l',U}$ or $T_w^{l',U}$, plot (b) in Fig. 4 demonstrates that, for a fixed $cx_s^{l'}/c_w^{l'}$ value, the execution delay remains constant for a $d_s^{l'}/d_w^{l'}$ ratio value of around 1, as it is dependent on the larger data size, $d_s^{l'}$. However, as the offloaded data size ratio surpasses 1, the execution delay increases for different values of $cx_s^{l'}/c_w^{l'}$.

Figure 5 shows the execution delay experienced by the users in a NOMA cluster for four distinct cases; no optimization, optimal power with equal cores, optimal power with random cores, and optimal power with optimal cores. In optimal power with equal cores assignment, power allocation is optimized, however, cores allocation is equal, i.e., $n_{c_1} = n_{c_2}$. In optimal power with random cores assignment, power allocation is optimized but the

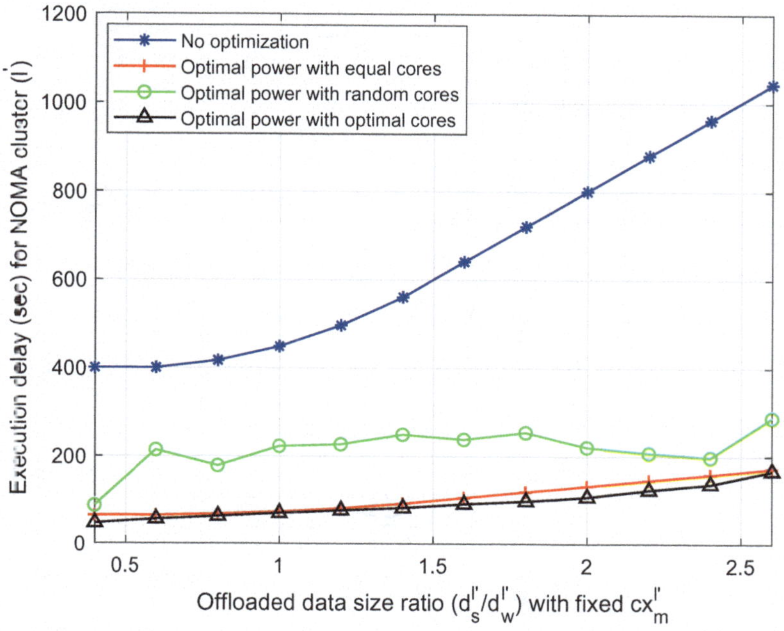

Fig. 5 Execution delay for four cases

cores allocation is random such that $\eta_s^{l'} = \alpha\, C_t$ and $\eta_w^{l'} = (1-\alpha)\, C_t$, where α is a random distribution uniformly varying between 0 and 1. The optimal power with optimal cores assignment case shows result for both power and cores optimization. All the results are attained by varying the offloaded data size ratio while the complexity is kept constant. It is observed that the execution delay is quiet high without any optimization. Notice that the optimal power assignment has a major contribution in reducing the execution delay in comparison to optimal cores assignment. It is also observed that the execution delay with equal cores assignment is lesser in comparison to random cores assignment because equal assignment of cores leads to a more effective distribution of resources, resulting in a reduction in execution delay in comparison to randomly assigned cores. The proposed scheme with optimal power and cores assignment achieves the minimum execution delay.

As already elaborated, a NOMA cluster cannot start a new transmission and computation process until the user having more computationally complex or bulk data completes its current transmission and computation. A higher value of execution delay difference among the paired users indicates bandwidth wastage. This difference is minimized by optimal power and cores allocation as shown in Fig. 6b. It is evident that the execution delay difference in plot (b) is greatly reduced in comparison to the preceding plot with no optimization. Furthermore, it is apparent that paired users attain optimal values of $d_s^{l'}/d_w^{l'}$ and $cx_s^{l'}/cx_w^{l'}$, that result in minimal difference in execution delay. For example, in plot (b) of Fig. 6, when $d_s^{l'}/d_w^{l'} = 1$ and $cx_s^{l'}/cx_w^{l'} = 1$, there is an execution delay difference of only 2 s. However,

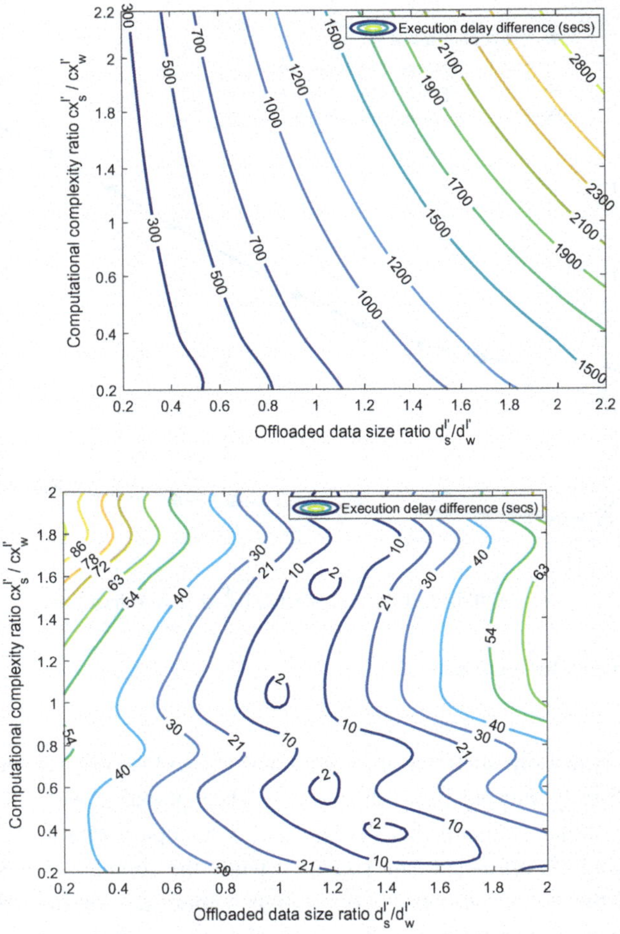

Fig. 6 Execution delay difference for no optimization & power and cores optimization

when $d_s^{l'}/d_w^{l'}$ increases to a value of 1.6 while maintaining the same $cx_s^{l'}/cx_w^{l'}$ ratio, the execution delay difference escalates to 36 s. Similarly, reducing $d_s^{l'}/d_w^{l'}$ to 0.7 results in an increase in the execution delay difference to 27 s. Thus, it is quite evident that the execution delay difference is minimized when the users having similar amount and complexity of offloading data are paired together which ultimately results in more efficient utilization of allocated resources.

Figure 7 shows the execution delay difference for the four distinct cases defined previously. All the results are attained by varying the offloaded data size ratio while the complexity is kept constant. The execution delay difference shows an increasing trend with increase in data size disparity among the paired users and is quiet high without any optimization. Power optimization greatly reduces the execution delay difference among the paired users. The

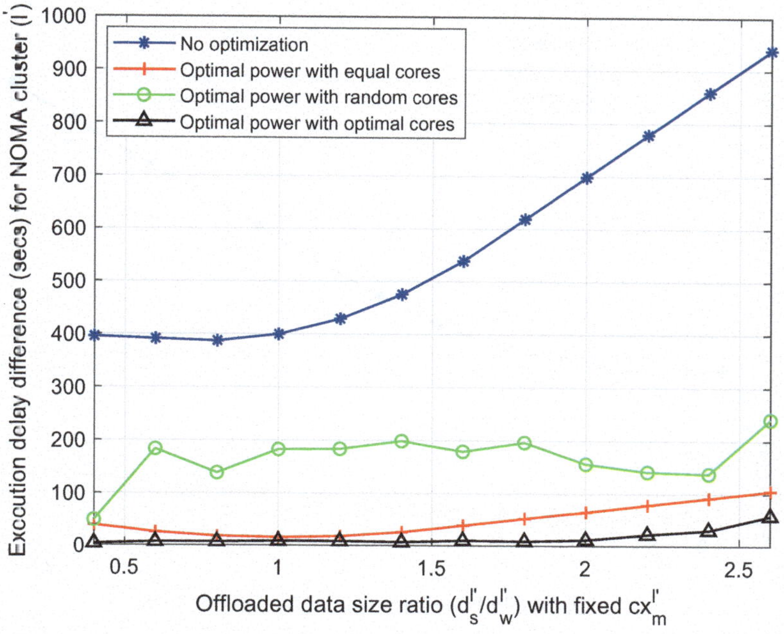

Fig. 7 Execution delay difference for four cases

execution delay difference is much reduced with equal cores assignment in comparison to random cores assignment as the former leads to more effective distribution of computational resources. The proposed scheme with optimal power and cores assignment achieves the minimum execution delay difference indicating minimal bandwidth and computational resource wastage.

To prove the authenticity of the presented optimization model, the problem of optimal power and cores assignment is subjected to a heuristic search across the entire solution space, referred to as “Simulation". The simulation results are then compared with the results achieved for optimal power and cores assignment using the Lagrange multipliers method, referred to as “Optimization" as shown in Fig. 8. The “No Optimization" case yields maximum execution delay in comparison to optimal power and cores assignment implemented in optimization and simulation results. Additionally, the convergence of simulation and optimization results reinforces the validity of our proposed optimization model.

Figure 9 shows a comparison of simulation and optimization results with respect to both power and cores assignment. In this result, data bits offloaded by weak user are kept constant ($d_w^{l'}$ = 4000 Kbits) while $d_s^{l'}$ are varied to obtain results. Consider the simulation results for optimal power assignment case shown in Fig. 9a, the allocated power to $T_s^{l',U}$ i.e., $p_s^{l'}$ increases with an increase in offloaded bits while $p_w^{l'}$ gradually decreases because of lesser bits offloaded by $T_w^{l',U}$ in comparison to those offloaded by $T_s^{l',U}$. On the other hand, the

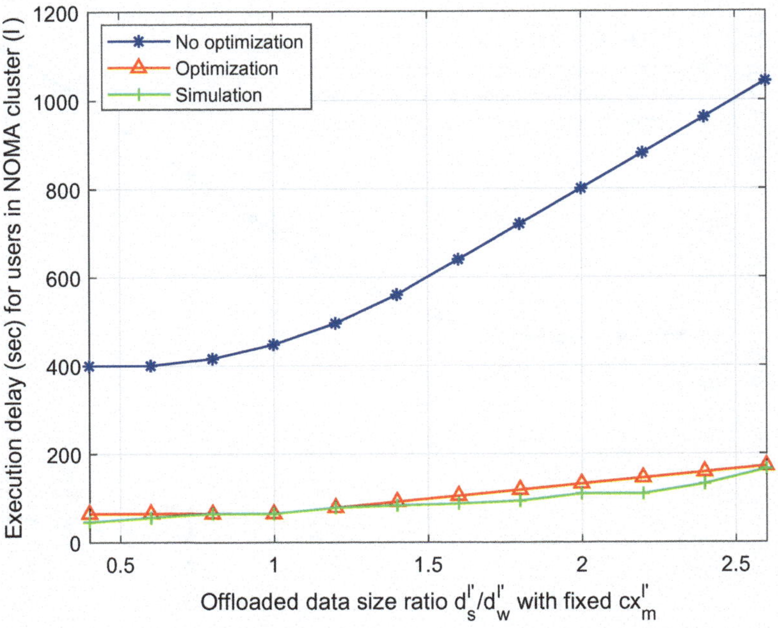

Fig. 8 Comparison of optimization and simulation results

values of $p_s^{l'}$ and $p_w^{l'}$ almost remain the same irrespective of offloaded data size in case of optimization results. While in Fig. 9b, the simulation and optimization results of cores assignment show a similar trend where a user with lesser offloaded data is assigned lesser cores and vice versa.

In the following analysis, we demonstrate the impact of minimal execution delay on the effective throughput, τ_{eff} defined as

$$\tau_{eff} = \frac{R_s^{l',U} + R_w^{l',U}}{\max(D_{e_s}^{l'}, D_{e_w}^{l'})}. \tag{40}$$

As indicated, τ_{eff} is the ratio of aggregated data rate achieved by the users in a NOMA cluster l' and the maximum of execution delay experienced by them. The resources allocated to users in a cluster are available for another execution process only when both the users complete their current execution process involving both the transmission and computation, hence the denominator incorporates the max(.) operator. It is noticeable that reduced execution delay enhances the effective throughput as shown in Fig. 10. Also, it is observed that a larger

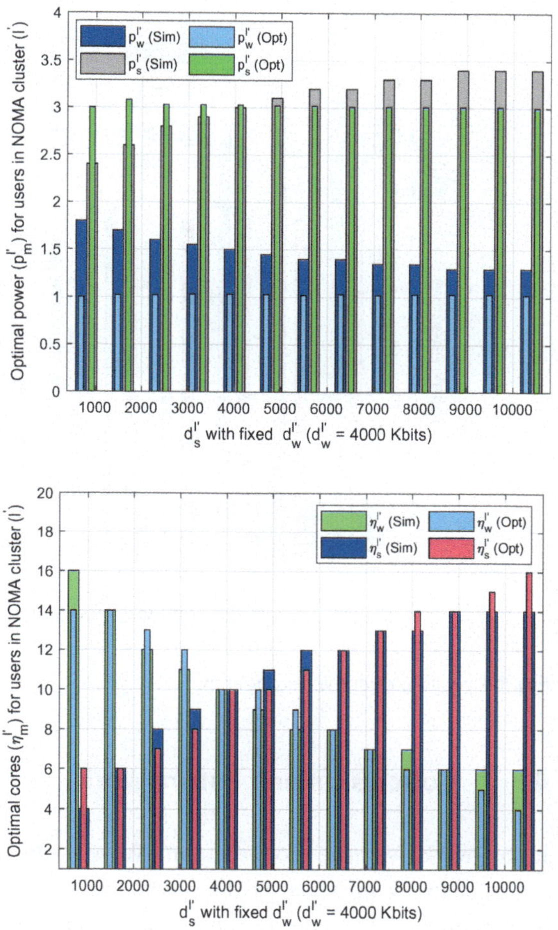

Fig. 9 Comparison of optimization and simulation results for optimal power and cores assignment

disparity in offloaded data size results in a greater difference in effective throughput among the compared cases. Furthermore, it is apparent that for fixed computational complexity and a range of $d_s^{l'}/d_w^{l'}$ values, the proposed power and cores optimization case outperforms other cases in terms of effective throughput. This trend is attributed to optimal core allocation which results in minimal execution delay in comparison to other cases. Additionally, it is observed that when the offloaded data exhibits the same characteristics with reference to amount and complexity, an equal number of cores are allocated for both equal cores and optimal cores assignment cases and as a result their throughput is quite similar with minor difference arising from the difference in power allocation for the two cases.

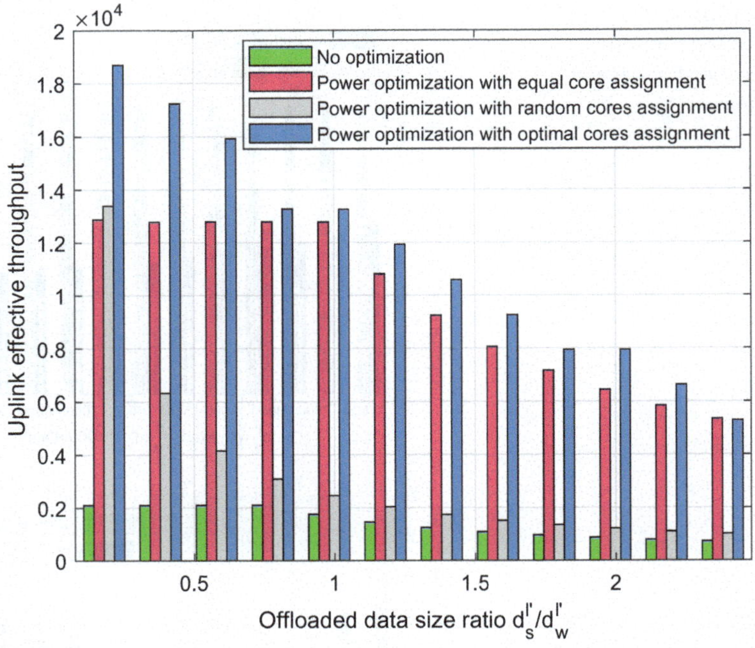

Fig. 10 System effective throughput

4 Conclusion and Future Work

In this work, we have proposed a mmWave enabled NOMA-based access and edge computing network in which HAPs provide access and edge computing services to ground user terminals. The mmWave band offers huge bandwidth which significantly increases achievable capacity in downlink and effective throughput in uplink. Also, NOMA due to its added benefits further contributes towards overall performance improvement. We devised a dual-layer optimization scheme for minimizing the execution delay. The transmission power and computational resources governing the execution delay are optimized by using Lagrange multipliers method and then convergence is attained by implementing a subgradient approach. Our study demonstrates the significant impact of minimized execution delay on overall resource utilization. Extensive simulation results validate our proposed solution, proving that in addition to providing access and edge computing services to user terminals in any remote area, the presented model is also able to reduce the overall computation and communication overhead and minimizes the inefficiencies in frequency and computational resource usage. As a forthcoming prospect, the model can be extended by considering multiple users in a NOMA cluster along with the consideration of a user's data attributes while clustering.

References

1. F. Babich, M. Comisso, A. Cuttin, M. Marchese, and F. Patrone, "Nanosatellite-5G integration in the millimeter wave domain: A full top-down approach," *IEEE Transactions on Mobile Computing*, vol. 19, no. 2, pp. 390–404, 2019.
2. M. Giordani and M. Zorzi, "Non-terrestrial networks in the 6G era: Challenges and opportunities," *IEEE Network*, vol. 35, no. 2, pp. 244–251, 2020.
3. M. Ke, Z. Gao, Y. Huang, G. Ding, D. W. K. Ng, Q. Wu, and J. Zhang, "An edge computing paradigm for massive IoT connectivity over high-altitude platform networks," *IEEE Wireless Communications*, vol. 28, no. 5, pp. 102–109, 2021.
4. M. Jia, L. Zhang, J. Wu, S. Meng, and Q. Guo, "Collaborative Satellite-Terrestrial Edge Computing Network for Everyone-Centric Customized Services," *IEEE Network*, 2022.
5. Y. Qiu, J. Niu, X. Zhu, K. Zhu, Y. Yao, B. Ren, and T. Ren, "Mobile edge computing in space-air-ground integrated networks: Architectures, key technologies and challenges," *Journal of Sensor and Actuator Networks*, vol. 11, no. 4, p. 57, 2022.
6. Q.-V. Pham, R. Ruby, F. Fang, D. C. Nguyen, Z. Yang, M. Le, Z. Ding, and W.-J. Hwang, "Aerial computing: A new computing paradigm, applications, and challenges," *IEEE Internet of Things Journal*, vol. 9, no. 11, pp. 8339–8363, 2022.
7. Q. Zhang, Y. Luo, H. Jiang, and K. Zhang, "Aerial Edge Computing: A Survey," *IEEE Internet of Things Journal*, 2023.
8. M. Latva-aho, K. Leppänen, F. Clazzer, and A. Munari, "Key drivers and research challenges for 6G ubiquitous wireless intelligence," 2020.
9. W. Jiang, B. Han, M. A. Habibi, and H. D. Schotten, "The road towards 6G: A comprehensive survey" *IEEE Open Journal of the Communications Society*, vol. 2, pp. 334–366, 2021.
10. 3GPP, "Study on New Radio (NR) to support non terrestrial networks," *TR 38.811*, no. Release 15, 2018.
11. N. Cheng, W. Xu, W. Shi, Y. Zhou, N. Lu, H. Zhou, and X. Shen, "Air-ground integrated mobile edge networks: Architecture, challenges, and opportunities," *IEEE Communications Magazine*, vol. 56, no. 8, pp. 26–32, 2018.
12. Z. Zhang, W. Zhang, and F.-H. Tseng, "Satellite mobile edge computing: Improving QoS of high-speed satellite-terrestrial networks using edge computing techniques," *IEEE network*, vol. 33, no. 1, pp. 70–76, 2019.
13. W. Wu, F. Zhou, B. Wang, Q. Wu, C. Dong, and R. Q. Hu, "Unmanned aerial vehicle swarm-enabled edge computing: Potentials, promising technologies, and challenges," *IEEE Wireless Communications*, vol. 29, no. 4, pp. 78–85, 2022.
14. M. S. Ali, H. Tabassum, and E. Hossain, "Dynamic user clustering and power allocation for uplink and downlink non-orthogonal multiple access (NOMA) systems," *IEEE access*, vol. 4, pp. 6325–6343, 2016.
15. L. Wang, Y. Wu, H. Zhang, S. Choi, and V. C. Leung, "Resource allocation for NOMA based space-terrestrial satellite networks," *IEEE Transactions on Wireless Communications*, vol. 20, no. 2, pp. 1065–1075, 2020.
16. Y. Wu, K. Ni, C. Zhang, L. P. Qian, and D. H. Tsang, "NOMA-assisted multi-access mobile edge computing: A joint optimization of computation offloading and time allocation," *IEEE Transactions on Vehicular Technology*, vol. 67, no. 12, pp. 12 244–12 258, 2018.
17. M. Giordani and M. Zorzi, "Satellite communication at millimeter waves: A key enabler of the 6G era," in *2020 International Conference on Computing, Networking and Communications (ICNC)*. IEEE, 2020, pp. 383–388.

18. D. Wang, M. Giordani, M.-S. Alouini, and M. Zorzi, "The potential of multilayered hierarchical nonterrestrial networks for 6G: A comparative analysis among networking architectures," *IEEE Vehicular Technology Magazine*, vol. 16, no. 3, pp. 99–107, 2021.
19. A. Umar, S. Basharat, S. A. Hassan, and H. Jung, "On the performance of multi-tier space-air-ground integrated network exploiting mmWave and THz capabilities for 6G communication," in *Proceedings of the 5th International ACM Mobicom Workshop on Drone Assisted Wireless Communications for 5G and Beyond*, 2022, pp. 49–54.
20. J. von Mankowski, E. Durmaz, A. Papa, H. Vijayaraghavan, and W. Kellerer, "Aerial-Aided Multi-Access Edge Computing: Dynamic and Joint Optimization of Task and Service Placement and Routing in Multi-Layer Networks," *IEEE Transactions on Aerospace and Electronic Systems*, 2022.
21. Y. Zhang, C. Chen, L. Liu, D. Lan, H. Jiang, and S. Wan, "Aerial edge computing on orbit: A task offloading and allocation scheme," *IEEE Transactions on Network Science and Engineering*, vol. 10, no. 1, pp. 275–285, 2022.
22. Q. Tang, Z. Fei, B. Li, and Z. Han, "Computation offloading in LEO satellite networks with hybrid cloud and edge computing," *IEEE Internet of Things Journal*, vol. 8, no. 11, pp. 9164–9176, 2021.
23. L. Cheng, G. Feng, Y. Sun, M. Liu, and S. Qin, "Dynamic Computation offloading in satellite edge computing," in *ICC 2022-IEEE International Conference on Communications*. IEEE, 2022, pp. 4721–4726.
24. N. Cheng, F. Lyu, W. Quan, C. Zhou, H. He, W. Shi, and X. Shen, "Space/aerial-assisted computing offloading for IoT applications: A learning-based approach," *IEEE Journal on Selected Areas in Communications*, vol. 37, no. 5, pp. 1117–1129, 2019.
25. Y. Wang, J. Zhang, X. Zhang, P. Wang, and L. Liu, "A computation offloading strategy in satellite terrestrial networks with double edge computing," in *2018 IEEE international conference on communication systems (ICCS)*. IEEE, 2018, pp. 450–455.
26. Y. Wang, J. Yang, X. Guo, and Z. Qu, "A game-theoretic approach to computation offloading in satellite edge computing," *IEEE Access*, vol. 8, pp. 12 510–12 520, 2019.
27. C. Qiu, H. Yao, F. R. Yu, F. Xu, and C. Zhao, "Deep Q-learning aided networking, caching, and computing resources allocation in software-defined satellite-terrestrial networks," *IEEE Transactions on Vehicular Technology*, vol. 68, no. 6, pp. 5871–5883, 2019.
28. S. R. Sabuj, D. K. P. Asiedu, K.-J. Lee, and H.-S. Jo, "Delay optimization in mobile edge computing: Cognitive UAV-assisted eMBB and mMTC services," *IEEE Transactions on Cognitive Communications and Networking*, vol. 8, no. 2, pp. 1019–1033, 2022.
29. A. Traspadini, M. Giordani, and M. Zorzi, "UAV/HAP-assisted vehicular edge computing in 6G: Where and what to offload?" in *2022 Joint European Conference on Networks and Communications & 6G Summit (EuCNC/6G Summit)*. IEEE, 2022, pp. 178–183.
30. S. Jeong, O. Simeone, and J. Kang, "Mobile edge computing via a UAV-mounted cloudlet: Optimization of bit allocation and path planning," *IEEE Transactions on Vehicular Technology*, vol. 67, no. 3, pp. 2049–2063, 2017.
31. F. Zhou, Y. Wu, R. Q. Hu, and Y. Qian, "Computation rate maximization in UAV-enabled wireless-powered mobile-edge computing systems," *IEEE Journal on Selected Areas in Communications*, vol. 36, no. 9, pp. 1927–1941, 2018.
32. W. Chen, B. Liu, H. Huang, S. Guo, and Z. Zheng, "When UAV swarm meets edge-cloud computing: The QoS perspective," *IEEE Network*, vol. 33, no. 2, pp. 36–43, 2019.
33. X. Hu, K.-K. Wong, K. Yang, and Z. Zheng, "UAV-assisted relaying and edge computing: Scheduling and trajectory optimization," *IEEE Transactions on Wireless Communications*, vol. 18, no. 10, pp. 4738–4752, 2019.
34. S. P. Boyd and L. Vandenberghe, *Convex optimization*. Cambridge university press, 2004.

Quality of Service Experimental Measurement for Satellite Communications Aircraft Terminal

Devika Divakaran, Ali Ali, Saba Al-Rubaye, and Antonios Tsourdos

Abstract

The growing demand for onboard services has provoked industry and academia to propose stable aeronautical networks that can overcome the significant link variations during aircraft long routes. Accordingly, the main focus of this chapter is employing, for the first time, the top-flight satcom (TFS) rack to measure the quality of service (QoS) metrics of real-time communications between suggested passengers and ground via satellite communications. For this purpose, the call traffic management algorithm is adopted to manage call/data transfer to control the response time of voice and data requests. Due to complexity and safety concerns related to making real-time onboard connections via satellites, a ground TFS testbed will be utilized to support the measurement. Measured results for power budgeting and QoS parameters are provided in the course of the chapter.

1 Introduction

Internet usage has witnessed an unprecedented surge in the last decade, which can be reflected in the number of emerging internet-based industries. The aviation industry is an example of an open market between terrestrial and satellite operators. Thus, considerable investments and new technologies have been proposed to enhance aeronautical networks [1]. However, the expected growth in the number of passengers in the upcoming years (around 19.3 billion in 2041 [2]) will probably lead to an increase in the burden on aeronautical networks as pertinent to their need to access internet services regardless of time and location [3]. In the same way, increasing the traffic load will lead to ultra-dense networks within airports [4].

D. Divakaran · A. Ali · S. Al-Rubaye (✉) · A. Tsourdos
School of Aerospace, Transport and Manufacturing, Cranfield University, Bedford, UK
e-mail: S.Alrubaye@cranfield.ac.uk

© The Author(s), under exclusive license to Springer Nature Switzerland AG 2025

M. A. Jamshed and A. Nauman (eds.), *Integrated Terrestrial and Non-Terrestrial Networks*,
Synthesis Lectures on Communications,
https://doi.org/10.1007/978-3-031-63705-6_6

Due to the limitations of using narrow beams and tilting angles of ground base station infrastructure, the fourth and fifth-generation networks (4G) and (5G) cannot support aircraft at high altitudes. That returns to the fact these base stations are mainly designed to serve customers distributed over the horizon rather than providing vertical coverage. Hence, the vertical coverage is limited to a few hundred meters only [5], which can be relatively enough for low-altitude platforms (LAP). Knowing that current rigorous efforts are towards extending that to support unmanned aerial vehicles (UAV) in the next generation of cellular communications [6]. Since the geostationary satellite (GEO) satellite is capable of maintaining a consistent connection with the aircraft regardless of where it is located in the satellite's footprint, continuous and uninterrupted communications are possible to be achieved between the satellite and the aircraft in the course of flight.

2 Related Work

Notable assets and the latest technologies have been proposed to support aeronautical networking development to allow sustainable, secure, and reliable connectivity. That can be witnessed in the stringent strives of the international civil aviation organization (ICAO) to achieve a higher scope of the global civil aviation system. Those efforts are shared with other entities in europe represented by the european organisation for safety of air navigation (EUROCONTROL) and the federal aviation administration (FAA) for the United States. Based on these prominent players, the aeronautical networks can be split into A2G network, AANETs, and satellite connectivity.

Although various companies have already deployed satellite constellations and ground stations to support aeronautical networks (Deutsche Telekom and Inmarsat through the european aviation network (EAN), Gogo Inc., and Collins Aerospace), the literature research has mainly investigated the aeronautical telecommunication networks *theoretically*. QoS parameters can be quite influential in evaluating the communication system performance during flight, so unpleasant scenarios can be avoided using the associate mitigation techniques or by utilizing backup links [7–9]. Satellite service providers are always keen on meeting the required QoS for aeronautical networks in terms of jitter, latency, and bandwidth. Efforts have been spent on network topology [10], aware traffic management [11], the impact of civil aeronautical networks routing protocols [12], and the performance of proposed aeronautical network architecture [13]. Still, there is a scarcity of QoS studies for aircraft communications with real-time experimental results for aircraft-to-ground links [14–17]. That needs to involve discussing software and hardware used onboard, satellite link budget, ground receiver link budget, and other important QoS parameters. Since the current research trend is towards embracing 5G and sixth-generation (6G) networks to aircraft [18–20], providing experimental results for aircraft to ground via satellites can be crucial for the future integration plans.

In this chapter, our objective is to provide valuable insights into aircraft-to-ground communication links facilitated by the TFS rack. We will cover the following key aspects:

- Rack system (hardware and software): We'll delve into the technical details of the TFS rack, exploring both its hardware components and the software that enables seamless communication between aircraft and ground stations.
- Potential ground-supporting links: We'll discuss the various ways in which the TFS rack establishes connections with ground-based infrastructure. This could involve cellular networks, satellite links, or other communication channels.
- Experimental evaluation of QoS parameters: QoS is critical for reliable communication. We'll conduct experiments to assess QoS parameters in two specific scenarios:
  - Land-moving vehicles: to evaluate QoS during calls with fast-moving vehicles.
  - Flying vehicles (aircraft): to evaluate QoS during calls with airborne vehicles.

By examining these aspects, we aim to enhance our understanding of aircraft-to-ground links via the TFS rack.

3 System Overview

Compared to the limited coverage of air-to-ground (A2G) networks, satellite networks can offer ubiquitous one on land, sea, and air [21], so they have been given priority for supporting onboard applications. The onboard data flow can be shortcut by a number of procedures. First, passengers request data content through the external phased array antenna (PAA) located at the upper surface of aircraft, by transmitting electromagnetic waves to the satellite (uplink), which processes these requests and sends them to the proper ground station after amplifing the waves (downlink). Knowing that ground station receivers operate at the downlink frequency, so received uplink signal is converted and processed to restore the original sent data by performing any necessary error correction and transmitted to the Earth. Afterward, the ground station allows access to the needed content by resending signals to the assigned satellite, which exchanges data again with the aircraft via the phased array antenna. That can be attained by the low Earth orbit (LEO), medium Earth orbit (MEO), or GEO satellites. To guarantee reliable communication, the entire process necessitates considering the equipment alignment, atmospheric attenuation, and other operational concerns [22]. As illustrated in Fig. 1, the system consists of three main segments: ground segment (ground station or gateway), space segment (GEO satellite in the context of this work), and aeronautical segment (aircraft equipped with TFS rack).

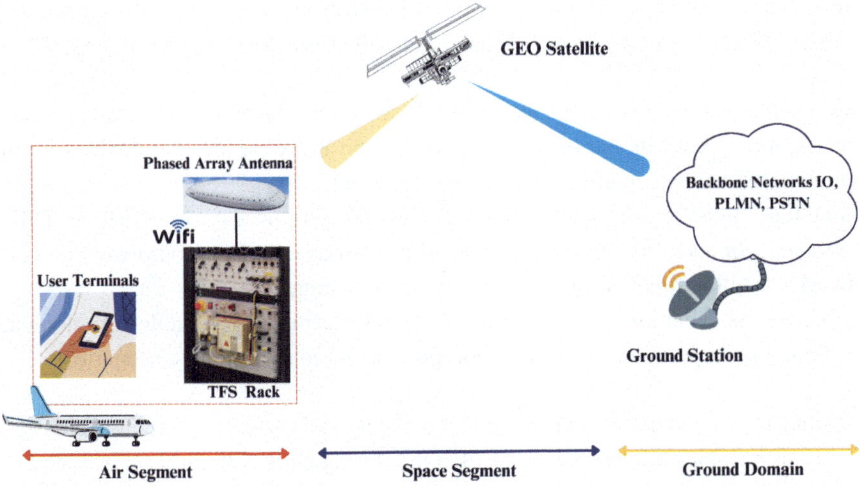

Fig. 1 Overview of main system equipment

4 Top-Flight SatCom Rack

The TFS rack allows passengers to reliably and securely enjoy sending/receiving e-mails, using social media platforms, during the flight, and even making calls through GEO satellites at the operator's frequency band. Figure 2a exhibits the main blocks of the TFS system, which usually encompasses line replaceable units (LRU) such as satellite data unit (SDU), SDU configuration module (SCM), diplexer low noise amplifier (DLNA), and intermediate gain antenna. As mentioned before, the TFS rack testbed will be employed in this study instead of the onboard TFS rack due to passengers' and flight safety concerns. The high-gain antenna (HGA) is located on building's roof, where it is Earth-bonded and lightning-protected. The antenna is followed by a DLNA and both the DLNA and HGA are linked to the SDU and SCM. However, the used TFS testbed can be shown in Fig. 2b, which is often mounted onboard and integrated with other avionics communication systems to guarantee reliable connections. The key subsystems of the TFS rack are:

1. Lock-out Unit: to stop any unauthorized usage/access to machinery or equipment
2. Audio Cockpit Panel: to manage and monitor audio/voice over communications
3. Draytek Wi-Fi Unit: to create a local network within the aircraft
4. Frequency Converter
5. Power Supply Unit
6. Power Consumable Unit
7. Server
8. Firewall Unit

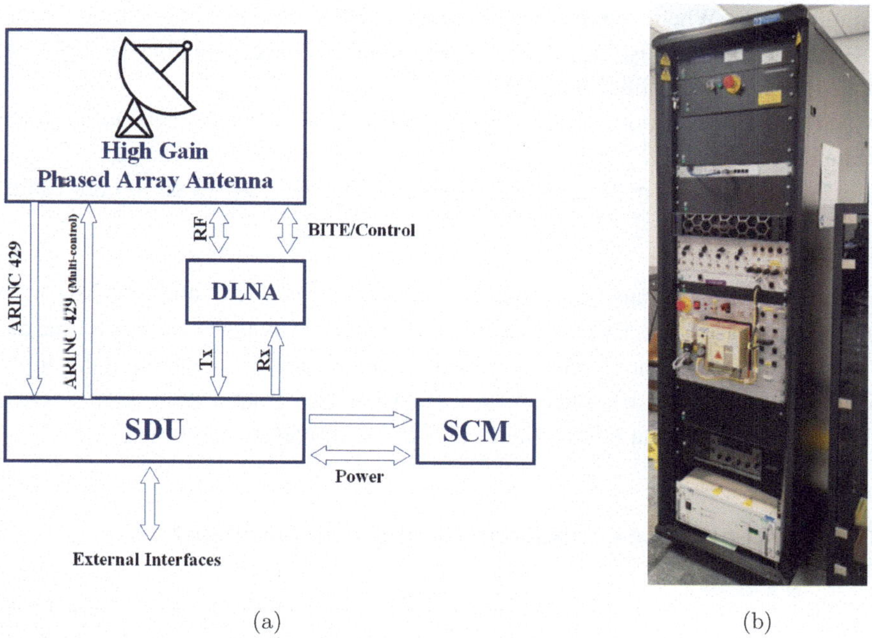

Fig. 2 TFS rack testbed. **a** Block diagram, **b** hardware

9. Satellite Data Unit: it will be discussed in the following subsection
10. SDU Configuration Module

4.1 Satellite Data Unit

The SDU plays a critical role in managing data flow and ensuring successful data transmission and reception over satellite links. As the central processing and control unit within the rack system, the SDU regulates operations related to SatCom connections. Its main functions include:

- Data Flow Regulation: The SDU ensures smooth data exchange between the aircraft and the ground, enhancing operational efficiency during flight.
- Dependable Connectivity: By enabling effective data connections, the SDU supports vital operations while the aircraft is in motion.
- Call Channel Management: It identifies and terminates call channels as needed.
- Interface Maintenance: The SDU handles interfaces, routing, and priority settings.

- Amplification: When equipped with an internal HPA, the SDU boosts radio frequency (RF) power. Optionally, an external flange-mounted power amplifier (FMPA) can be added for further amplification [23].
- Modulation, Error Correction, and Coding: The SDU manages these tasks to ensure seamless connectivity.
- Flexible Configuration: Its adaptable design allows the SDU to serve both cabin and cockpit services.

The employed SDU operates in the frequency ranges between 1525 and 1559 MHz (receiving) and 1626.5 and 1660.5 MHz (transmission), which are assigned to improve the quality of signals and decrease interference in pursuit of attaining efficient satcom links. It plays a critical role in promoting data-intensive voice and data-based applications, aircraft tracking, retrieval of meteorological information, and flight performance monitoring.

4.2 Multi-purpose Control and Display Unit Simulator

The multi-purpose control and display unit (MCDU) is an interface used to interact with the aircraft's flight management system (FMS) and to obtain different types of data, including link parameters. The MCDU can be in the form of hardware or software. In the case of using the former, a component will be attached to a computer. As for the software form, an ARINC 429 interface box is connected to the PC to execute the software program entries. The user-friendly MCDU simulator can be used to make calls as described below:

4.2.1 VoIP Communication Using MCDU

One of the key services supported by the TFS rack is making voice-over internet protocol (VoIP) phone calls with the ground for emergency cases. The configuration and dial windows are displayed in Fig. 3. The main dial window is illustrated in Fig. 3b, where it shows several parameters like link status and called numbers. A directory index is assigned for each phone number entered.

4.2.2 Voice and Data Transmission Management

Figure 4 exhibits the used algorithm for VoIP calls management that minimizes the blocking probability of data while optimizing QoS metrics. A transmission path needs to be assigned for each transmission, so packets can be serviced or buffered (queued) depending on the path availability or unavailability, respectively. This buffer management and call admission algorithm works on minimizing the probability of data blockage and optimizing the quality of service parameters. Provided that the transmission path was engaged, voice and data packets would be buffered, taking into account that voice packets will at all times enjoy higher priority over data packets [24]. Since data packets are less sensitive to delay in

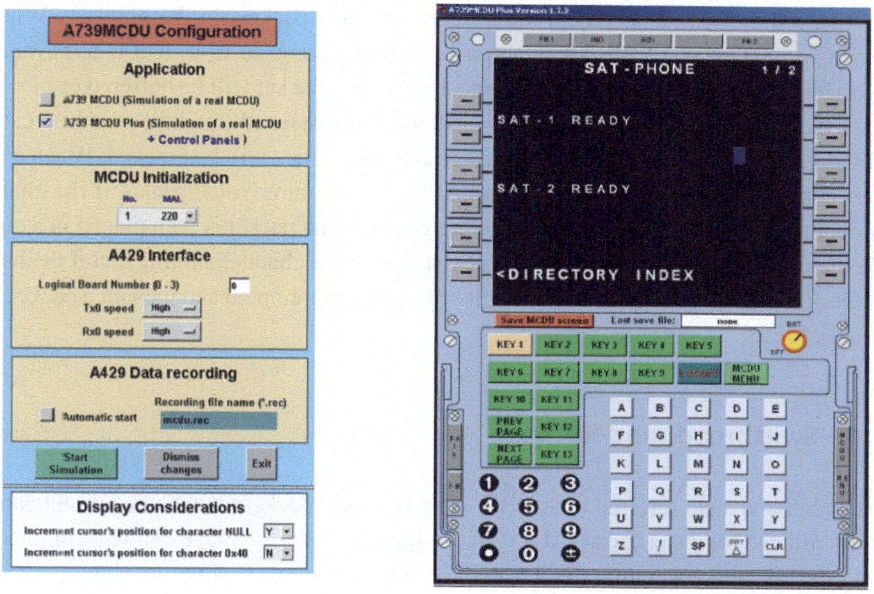

(a) (b)

Fig. 3 The MCDU tool. **a** Configuration window, **b** dial window

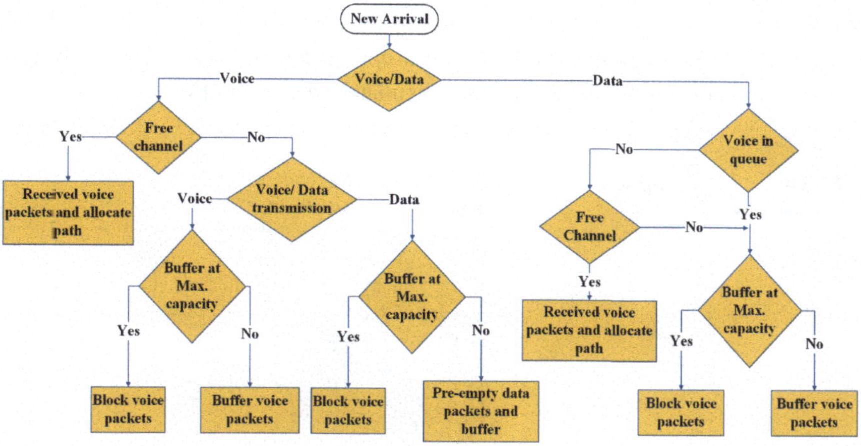

Fig. 4 The algorithm for call traffic management

most cases, they will be served in the absence of voice packets in the network. When a transmission path is available, so voice packets can be assigned to it. If it is not available due to serving another voice request, the incoming voice packet will be buffered or blocked. In case the buffer maximum predefined capacity is not reached, the voice packets can be buffered until approaching the maximum capacity where they will be blocked. When a data packet is being processed, it gets interrupted by an incoming voice packet if the channel isn't operating at maximum capacity. The pre-empted data packet is then placed in a queue while the voice packet takes priority and is assigned to the channel. Noting that if the buffer maximum capacity is attained, data packets will not be pre-empted and the voice packet will be blocked [25].

4.3 Swiftbroadband Network

SwiftBroadband, also referred to as SBB, is an IP-based packet-switched communications network utilized for the proposed TFS rack. It supports global coverage onboard commercial aircraft over the 4th generation satellites. SBB offers seamless coverage and communication by the network which utilizes satellites and ground stations. Hence, passengers enjoy using multiple programs to send emails, browse the internet, and make video calls with an internet speed of around 432 Kbps per onboard channel. Unlike charge-by-the-minute services, SwiftBroadband is always connected and ready to use. Passengers are not charged per minute, they are only charged for the amount of data transferred in blocks of Megabytes (MB) instead. The Asia-Pacific, Middle East, Europe, Americas, and Africa are the primary geographic areas covered by the SBB as approximately defined in Fig. 5.

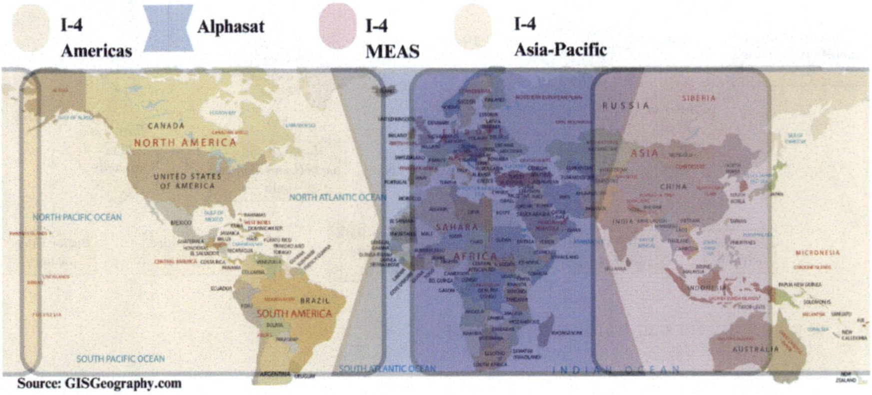

Fig. 5 SwiftBroadband coverage area

4.4 ARINC-781 Standard

Aeronautical Radio, Incorporated (ARINC) Corporation is a global leader in transport communications and systems engineering solutions. The airline electronic engineering committee (AEEC) establishes standards, which ARINC both copyrights and publishes. The AEEC, comprising major airline companies, avionics manufacturers, and ARINC members, defines the structure and function of avionics equipment through unique documents that describe their operation. The ARINC series of characteristics are categorized into 400, 500, 600, and 700. Notably, ARINC 781 serves as the industry standard for Inmarsat SatCom capability in Classic Aero and SwiftBroadband operations. The ARINC 781 SatCom Avionics System consists of the SDU, SCM, and FMPA [23].

5 Potential In-Flight Connectivity Links

The demand for improved in-flight experience is the primary motive behind the recent unmatched development of in-flight connectivity (IFC) services to follow the accelerated rhythm of application requirements. Over the past decade, the aviation industry has exerted significant pressure on cellular and satellite service providers to enhance global performance. Ensuring seamless wireless connectivity across the entire aircraft cabin is poised to become a central focus for future in-flight entertainment systems (IFE) [26]. Employing satellite links through high throughput satellites, A2G links supported by 4G and 5G networks, or using AANET links are the possible options to allow onboard communications with persistent enhanced service. Onboard aeronautical networks-assisted IFE services can be divided into passenger entertainment, inform passengers, and cabin crew communications [27].

5.1 Air to Ground Connectivity

A2G connectivity allows the exchange of data up and down between the aircraft and the ground, so passengers experience services quickly and smoothly onboard [28, 29]. It is worth mentioning that these services are not limited to passenger communications but also are employed for flight operations and air traffic management. The growing demand for connectivity puts an increased burden on the onboard internet service providers to reach a higher quality of experience (QoE) for passengers. Direct A2G (DA2G) and satellite A2G (SA2G) are the two categories for A2G links, which are illustrated in Fig. 6a and b, respectively. In the realm of modern aviation, both DA2G and SA2G systems play a crucial role. These systems facilitate real-time communication, enhance flight safety, and significantly elevate passenger comfort through improved In-Flight Connectivity [14].

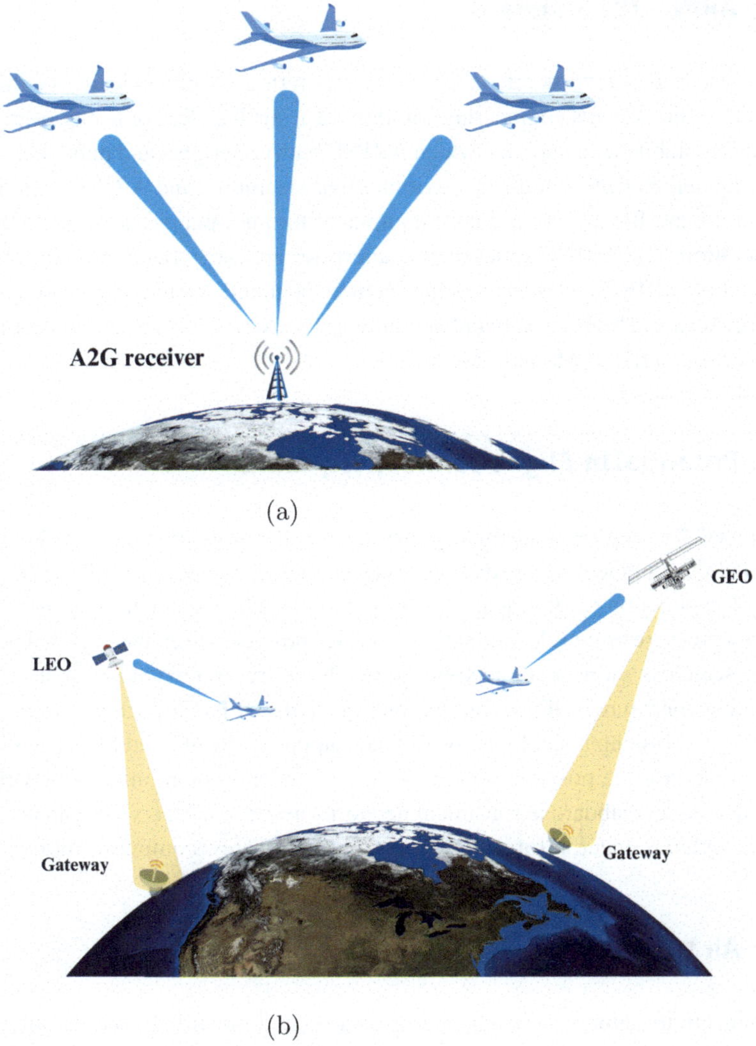

Fig. 6 A2G links. **a** DA2G, **b** SA2G

5.1.1 Direct A2G Connectivity

DA2G establishes a direct connection between the aircraft and ground stations. Terrestrial infrastructure-based communications are used to make connections between the aircraft and ground stations [30]. Typically, DA2G communications can be adopted for communications within a specific region, so real-time applications benefit from lower latency and higher data rates by means of these links. Continental flights, which constitute a significant portion of an airline's revenue, benefit greatly from the DA2G communication system. DA2G offers several advantages, including in-cabin long-term evolution (LTE) services, quick login, and high data rates. Users can rely on DA2G to maintain uninterrupted cellular connections (LTE and 5G). Unlike satellite A2G systems, which suffer from latency issues, DA2G is ideal for supporting applications with stringent QoS requirements, such as video calls and streaming.

Air-to-LTE (A2L) networks, as an example of DA2G communications, are a new form of connection that utilizes ground cellular base stations, deployed along the flight route, to allow passengers to enjoy real-time high-quality video streaming (up to 4K) and web browsing. These connections require expanding the vertical coverage of the route's ground cellular stations or adding energy-efficient antennas for that purpose. Currently, the LTE ground stations are capable only of supporting UAV communications [31, 32]. A2L networks, utilizing LTE technology, can achieve data rates of up to 75 Mbps between the aircraft and ground stations. Additionally, speeds of 25 Mbps are attainable when the aircraft is 100 km away from the ground station, traveling at a speed of 1200 km/h. However, maintaining line-of-sight (LOS) connectivity poses a challenge in A2L communications. Factors such as channel impairments, Doppler shift, handover issues, and interference may hinder the performance of current cellular networks. Therefore, utilizing technologies such as multi-steerable beam antennas, beamforming, and spatial multiplexing techniques can allow the simultaneous exchange of the desired data with various geographically dispersed aircraft.

5.1.2 Satellite A2G Connectivity

SA2G connections engage satellites as an additional middle step to connect aircraft to ground stations, which makes them the perfect choice for intercontinental long-distance flights in the absence of terrestrial infrastructure. While A2L networks can achieve impressive data rates of up to 75 Mbps between the aircraft and ground stations using LTE technology, they may suffer from higher latency and reduced speeds due to the vast distances involved. Despite these challenges, a hybrid network that combines DA2G and SA2G communication is seen as a viable solution for providing a comprehensive A2G connection [14].

Emerging satellite technologies are poised to revolutionize IFC by enveloping the globe with high-powered, wide-bandwidth coverage. These advancements will enable seamless connectivity for devices on both commercial and business aircraft, serving millions of passengers. The three primary types of satellites employed for this purpose are LEO, MEO, and GEO. LEO satellites orbit relatively close to Earth, typically at altitudes ranging from 500 to 2000 km above the Earth's surface [33]. The round-trip time (RTT) for these satellites is approximately 30 ms. LEO satellites move so fast across the sky, orbiting at speeds of 7.8 km/s, and therefore require a lot of effort to be tracked from ground stations. Thus, the time of connection lasts for a few minutes between 15 and 20 min. MEO satellites are orbiting at higher altitudes ranging from 5000 to 20,000 km, which leads to higher latency with an RTT of about 100 ms. Whereas GEO satellites are positioned on orbits of 35,786 km above the Earth's surface, so they have an RTT of 250 milliseconds and have the highest latency. GEO satellites have a wider coverage area and higher capacity compared to LEO and MEO satellites. Although L, S, and Ku bands currently dominate the deployed GEO satellites, they suffer from limited bandwidth. Accordingly, high-frequency bands such as the Ka and Q/V bands have been recently adopted in different satellite constellations to meet the increased demand for internet services [34]. Knowing that satellite links can be based on RF or optical communications, as below:

- RF Communications: an RF signal is a wireless electromagnetic signal used as a form of communication through radio waves with defined frequencies in the range of 3 kHz to 300 GHz. RF propagation occurs at the speed of light. These waves are employed in television broadcasting, radar systems, remote control, remote metering/monitoring, and others. The chosen frequency band is adopted based on the application's requirements as higher frequency bands offer broader bandwidth to support higher data rates [35]. However, utilizing higher frequency bands leads to increasing atmospheric impairments due to rain, gas, and other weather conditions.
- Optical communications: it involves transmitting signals using light rather than electrical current. These systems rely on optical mediums and consist of several components, including modulators/demodulators, transmitters/receivers, light signals, and transparent channels. As an emerging technology, optical communications play a crucial role in satellite-to-aircraft links that utilize LoS signals. In optical networks, data is carried using lasers, infrared, or near-infrared waves as an alternative to traditional RF signals. Compared to RF, optical communications offer significantly higher data rates, often exceeding hundreds of gigabits per second. This makes them an attractive choice for efficient airplane communication, especially considering their lower power requirements and smaller antenna size. Optical beams are particularly well-suited for establishing backbone connections between aircraft and ground stations. However, the quality of received signals and connection performance can be affected by atmospheric conditions. For links below 1.2 km, fog and low clouds pose primary concerns, while scintillation, heavy rain, and snow can significantly impact links exceeding 1.2 km [36].

5.2 Aeronautical Ad-Hoc Network

The AANET is envisaged to support broadband IFC through autonomous and self-configured wireless networks. The AANET uses multi-hop Air-to-Air (A2A) radio communications through aircraft, together with integrating satellite and ground cellular networks. As illustrated in Fig. 7, the AANETs mainly consist of satellite, aircraft (inter-links), and ground layer, respectively [37]. AANET belongs to the mobile ad hoc network (MANET) family, which was introduced in the 1990s to allow direct wireless connections among adjacent equipment without employing network infrastructure. AANET inherits some features of the MANET networks as the dynamic nature of topology, but AANET varies in terms of the

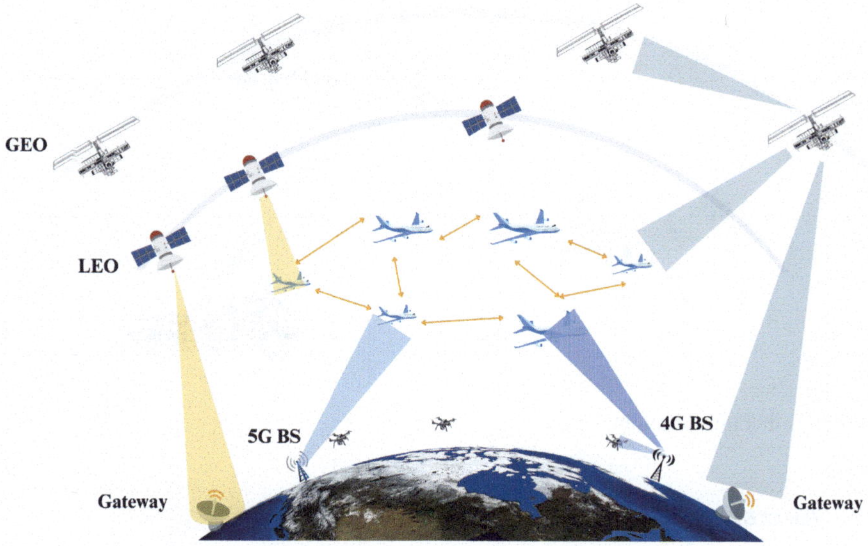

Fig. 7 The topology of AANET

Table 1 Comparison between satellite connectivity, A2G and AANETs

Aspect	Satellites Net.	A2L	AANET
Range	Global coverage	Proximity of BSs	Limited
Infrastructure	Space-based equipment	Ground-based equipment	Aircraft-based equipment
Latency	High	Low	Variable
Cost	High	Low	Moderate
Bandwidth	Relatively broad	Broad	Limited

density of the nodes, speed, and route of movement. Table 1 compares satellite, A2L, and AANET networks, as below:

6 Air to Vehicles Scenarios

In this section, two air-to-vehicle (A2V) scenarios will be proposed. The first one will concentrate on evaluating the communication system performance between the aircraft and a moving car. While aircraft-to-UAV communication system will be introduced and assessed in the second scenario.

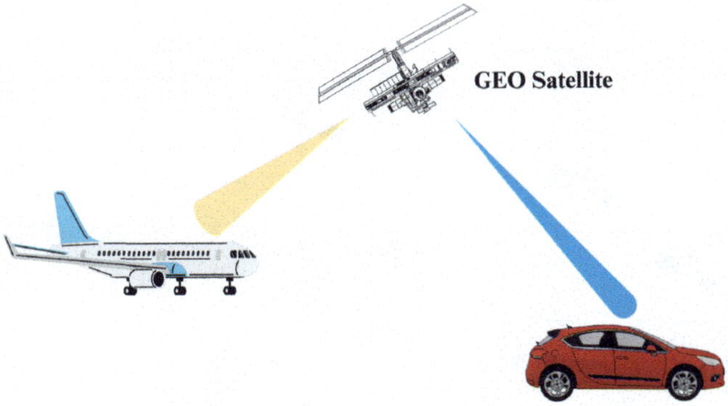

Fig. 8 SA2G link (aircraft to moving car)

6.1 Aircraft to Car Communication System

This scenario discusses the QoS parameters of SA2G links while making a voice call using the TFS rack between a passenger and a high-speed car, as illustrated in Fig. 8. The introduced scenario provides an insightful analysis of the link taking into account the recipient's mobility. It is worth mentioning that the QoS measurements will focus mainly on the links between the aircraft and the GEO satellite and between the car and used nearby base station. Therefore, the TFS simulator will be used for satellite links and the XCAL tool will be exploited for measuring the QoS for the terrestrial links.

To perform the measurement, the car followed different paths like the one shown in Fig. 9 based on the permitted speed limits. The path below, captured by the XCAL tool, contains different colors each referring to a specific network (blue: 3G, green: 4G network, and sky blue: 5G). Unfortunately, the nearby paths are not covered by a 5G network, so only 3G and 4G are displayed on the map.

6.1.1 System Configuration and Results

First, the TFS terminal was turned on through a sequence of steps, which can be summarized by powering the consumer unit, the frequency converter, and the server, respectively. After that, the SDU unit LEDs were green, which indicates a successful connection between the satellite and the TFS rack. Then, the rack was connected to the MCDU simulator, so that we were able to dial the desired number through the rack. As for the car side, the Accuver XCAL tool, installed on a Sony smartphone, is adopted to evaluate the required metrics. XCAL tool is a standalone application compatible with the Android operating system (OS) and provides reliable QoS and quality-of-experience (QoE) measurements. This tool offers various measurement AutoCall types (Voice, VoLTE, FTP, HTTP, SMS, email, ping, iPerf, YouTube, ARS-Incoming, Multi-Call, Multi- RAB, Multi-Session, Mobile SNS, Mobile

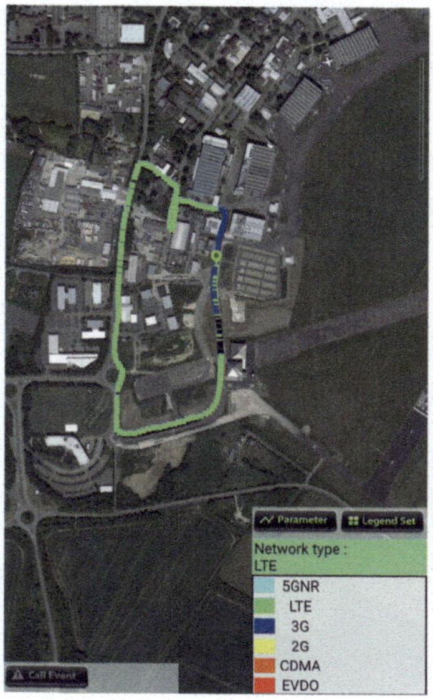

Fig. 9 Trajectory of one of the used roads

Cloud, ADB Script, Messenger Talk, and Messenger Texting.), so voice option was selected to complete the measurement.

Satellite Links:

This part will utilize the TFS simulator and maintenance menu to obtain satellite-to-ground link parameters. To analyze the power budgeting and QoS parameters of aircraft to satellite links, the specifications of the satellite and TFS rack's antenna need to be considered. The PAA of the testbed is an AMT-3500 intermediate gain antenna, which is $87.6 \times 24.1 \times 6.4$ cm. The operational temperature of the antenna is -57–$71°$ and EIRP equals 20 dBW. While the minimum gain is 6 dBic for 98% of the coverage area (minimum 10 dBic for 75% of the coverage area). Regarding the satellite specifications, the EIRP equals 42 dBW for a transmission gain of around 30 dB. These values and other parameters are used to calculate the free space path loss (FSPL) in clear weather on the measurement day as shown in Fig. 10a using the equation:

$$FSPL = 20\log(f) + 20\log(d) - 147.56 \tag{1}$$

where the frequency is f, d denotes the distance between transmitter and receiver. Accordingly, the carrier-to-noise ratio is given by:

$$\frac{C}{N}_{up/down} = EIRP_{up/down} - FSPL_{up/down} + 10\log(G_R) - 10\log(T) - 10\log(k) - 10\log(B) \tag{2}$$

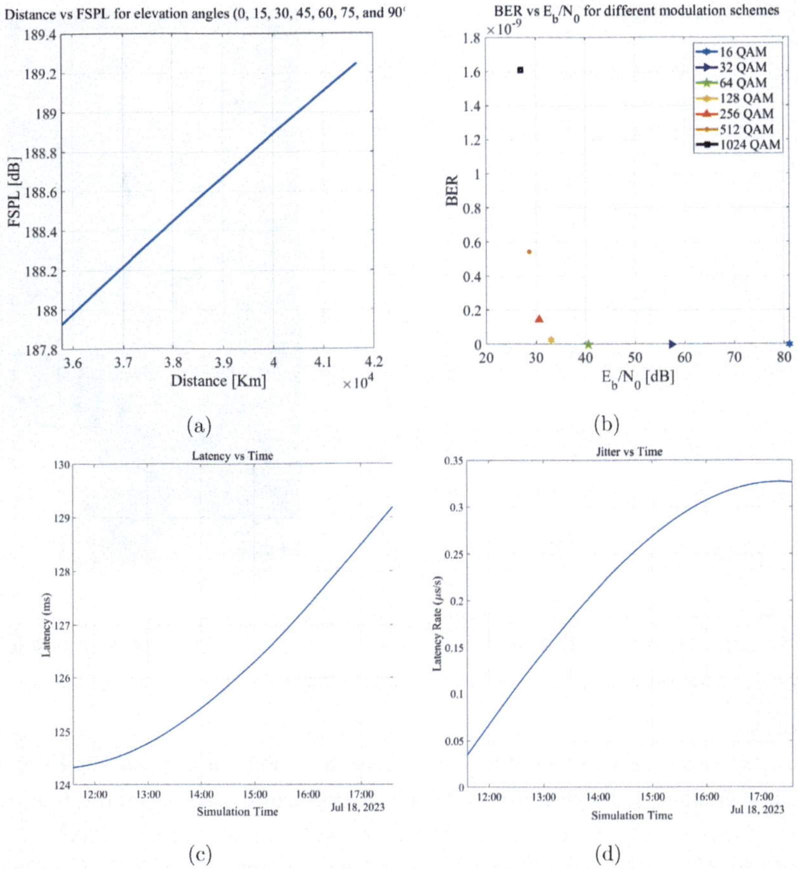

Fig. 10 Measured results. **a** FSPL for different satellite elevation angles, **b** BER for the associated modulation schemes, **c** latency, **d** jitter

where the receiver gain is G_R, T denotes receiver temperature, k refers to Boltzmann constant, and the noise bandwidth is B. The TFS rack can be connected with more than one satellite at the same time, so the FSPL is computed for different satellite elevation angles. The bit error rate (BER) is used to evaluate the performance of the up-/downlinks between the TFS rack and the satellite. Satellites employ adaptive coding and modulation (ACM) to enable dynamic changes in communication system throughput based on weather conditions, so several modulation schemes are examined in Fig. 10b. The BER can be calculated through:

$$\frac{E_b}{N_0} = \frac{C}{N} - 10\log(e^M * 0.5) \tag{3}$$

where M depends on the used modulation scheme.

Table 2 XCAL-based QoS measurement for SA2G link

Time	16.30	16.40	16.50	17.00
Speed (km/hr)	30	40	50	60
G1 Rx (dB)	−0.90	−58.50	−56.40	−54.80
G1 Tx (dB)	−18.80	−16.70	−14.40	−12.00
Max. Ec/Io (dB)	−3.13	−2.44	−2.51	−2.85
Max. RSCP (dBm)	−65.39	−63.47	−61.60	−59.69
Rx power (dBm)	−58.1	−56.3	−54.4	−52.8
Tx power (dBm)	−18.4	−16.9	−15.7	−14.0
CQI	24	22	12	16
DTX rate (%)	100	100	98.39	99

On the other hand, Fig. 10c and d portray the variations in latency and jitter values as functions of the distance between the PAA and the GEO satellite. However, more fluctuations would be observed when a flying TFS rack is considered.

Terrestrial Links:

The measurement campaign was performed by recording the QoS metrics during time slots at different speeds between 16.30 and 17.00 PM on the measurement day, as revealed in Table 2. The measured parameters are:

- Received gain of the first smartphone antenna (Rx).
- Transmitted gain of the first smartphone antenna (Tx).
- Received power.
- Transmitted power.
- Energy per chip over interference power ratio (Ec/Io): a metric utilized for the connection quality evaluation of the user's signal.
- Received signal code power (RSCP): the received signal strength of a user's connection. A stronger received signal will be indicated by a higher RSCP value, whereas a lower RSCP value will indicate a weaker signal.
- Channel quality indicator (CQI): a 4-bit value that indicates the highest modulation and code rate for a received transport block that meets a block error rate.
- Discontinuous transmissions (DTX): a method of momentarily powering down, or muting, a telephone set when there is no voice input to the set.

Figure 11 illustrates the parameters above at speeds of 30, 40, 50, and 60 km/hr. The movement duration at each speed was around ten minutes, which justifies the observed variations in the values.

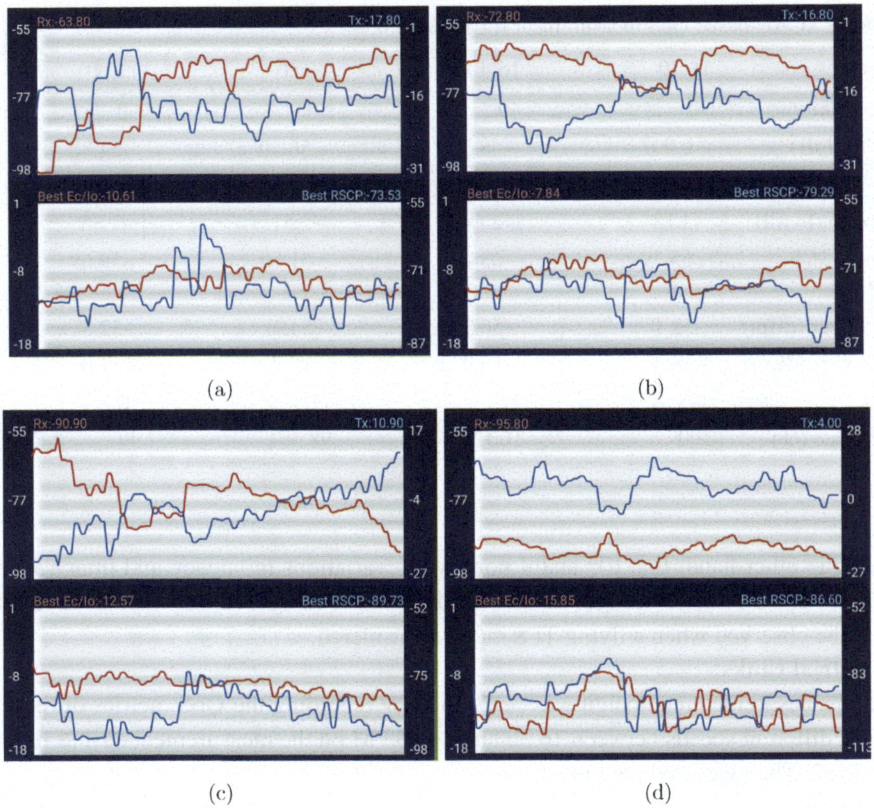

Fig. 11 Measured RSRP and Ec/Io, G1, G2 for the proposed scenario at different speeds. **a** 30 km/hr, **b** 40 km/hr, **c** 50 km/hr, **d** 60 km/hr

Since service times and arrival rates (number of packets) are required to analyze the latency and jitter of received packets (traffic flow), Table 3 will be used as it counts the received packets and provides timestamps and locations with a second time slot during the car movement. The statistics of the received packets can be given as follows all received packets (842), minimum received packets per timestamp (2), maximum received packets per timestamp (53), average received packets (18.76), standard deviation (13.73), and median (13). Accordingly, Fig. 12a and b show the latency and jitter of the received packets by the car rider's smartphone when a call is made using the TFS rack. Three nodes should be considered when it comes to analyzing calls made from aircraft to ground users through the TFS rack, which are aircraft, satellite, and ground systems. If d_u represents the average packet transmission delay for a link u, the average delay for the origin-destination (o, d) pair on their path p can be expressed by [38]:

Table 3 Couning the received packets via the XCAL tool

No.	Timestamps	Longitude	Latitude	Count of Rx packets
1	2023-07-26 17:11:30	−0.62993	52.06645	12
2	2023-07-26 17:11:31	−0.629925	52.06645	7
3	2023-07-26 17:11:32	−0.62992	52.066455	12
		⋮		
841	2023-07-26 17:25:34	−0.62833	52.06776	31
842	2023-07-26 17:25:35	−0.62833	52.06776	16

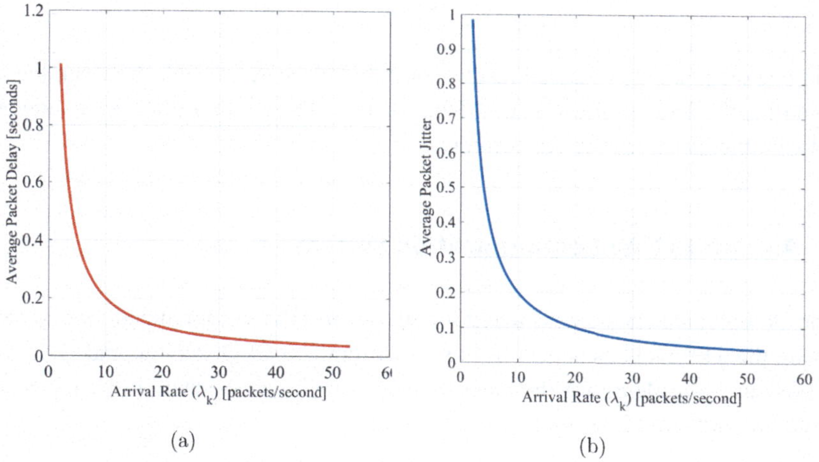

Fig. 12 Measured results. **a** Latency values for varied arrival rates, **b** jitter values for varied arrival rates

$$D_p^{o,d} = \sum_u A_{p,u}^{o,d}.d_u \tag{4}$$

where $A_{p,u}^{o,d}$ is the link-path incidence matrix. An element is equal to 1 if link u is in path p between pair (o, d) and zero otherwise.

Hence, the weighted average delay is:

$$D^{o,d} = \frac{1}{\lambda^{o,d}} \sum_u X_p^{o,d}.D_p^{o,d} \tag{5}$$

where $X_p^{o,d}$ stands for the amount of (o, d) traffic on path $p \in P^{o,d}$. Accordingly, The total weighted delay can be given by:

$$D = \frac{1}{\gamma} \sum_{o,d} \lambda^{o,d} . D^{o,d} \tag{6}$$

where γ refers to the total packet arrival rate in the network and λ is the total arrival rate.

By following the same steps, the path, commodity (origin to destination pair), and network jitter can be derived as:

$$J_p^{o,d} = \sum_u A_{p,u}^{o,d} . J_u \tag{7}$$

$$J^{o,d} = \frac{1}{\lambda^{o,d}} \sum_u X_p^{o,d} . J_p^{o,d} \tag{8}$$

Finally, the network jitter can be expressed by:

$$J = \frac{1}{\gamma} \sum_{o,d} \lambda^{o,d} . J^{o,d} \tag{9}$$

It should be noted that the obtained results consider only wireless links between the smartphone and nearby base stations. It is evident that the latency and jitter are exponentially reduced with the increase in the arrival rate.

6.2 Aircraft to UAV Communication System

Here, the QoS parameters are considered for aircraft-to-UAV communications (exchanging data) using the TFS rack, as shown in Fig. 13. Aircraft to GEO satellite and UAV to base station links are considered for the measurement, which was carried out by means of the TFS simulator and the XCAL tool.

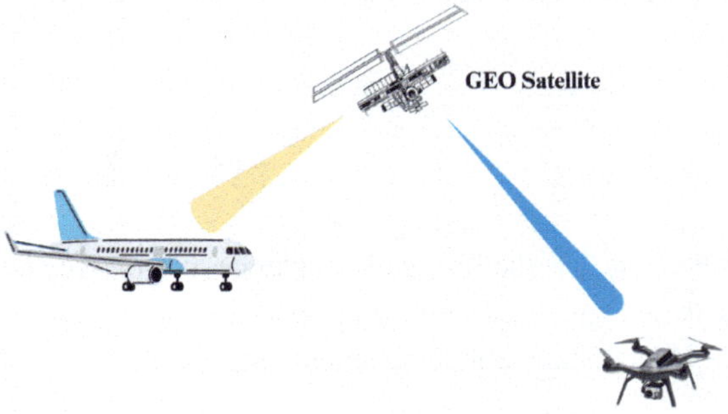

Fig. 13 SA2G link (aircraft to UAV)

6.2.1 System Configuration and Results

The UAV model and the flying site were chosen to be aligned with the UK civil aviation authority (CAA) regulations. Accordingly, for the same previous configuration for the TFS testbed, a call was made between the aircraft and the UAV, which hovered at a maximum altitude of 50 m as will be seen in the following:

Satellite Links:

The configurations of the TFS rack and satellite equipment did not change in the second scenario, so the previous satellite link results are valid for both scenarios.

Terrestrial Links:

The site to fly the drone was nominated to be covered mainly by a 5G network, so the obtained results were not only important for the aircraft to UAV scenario but also delivered thoughtful information regarding the performance of 5G networks at an altitude of 50 m. Whereas the accompanied signal-to-noise-and-interference ratio (SINR) can be outlined in Fig. 14a. Despite the strong 5G coverage in the selected area, Fig. 14b reveals that during the call reception from the TFS rack, LTE networks slightly contributed to the total throughput.

7 Conclusion

This chapter discussed the applicable links to support the IFC services, described the hardware and software used in the TFS rack to support aircraft to satellite communications, and proposed two scenarios to explain making onboard connections with land-moving and flying vehicles using the TFS rack. Also, the QoS parameters for the employed 4G, 5G, and satellite links were introduced. Further QoS investigations can be made using newer generations of the TFS rack, which embraces powerful antennas and adopts the latest techniques to attain a higher system performance. Moreover, with the accelerated research towards the 6G networks, it will be crucial to study how 6G links can support aircraft in A2G links, along with the 6G integration with satellite links.

Fig. 14 Measured results. **a** Path-loss of the 5G signals, **b** SINR, **c** throughput

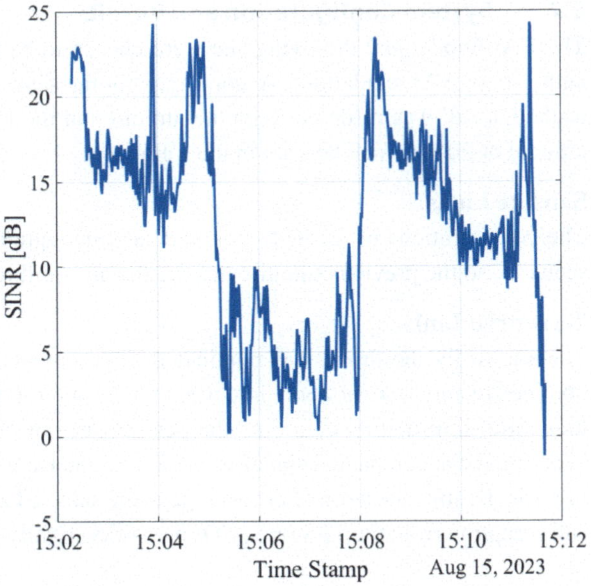

(a)

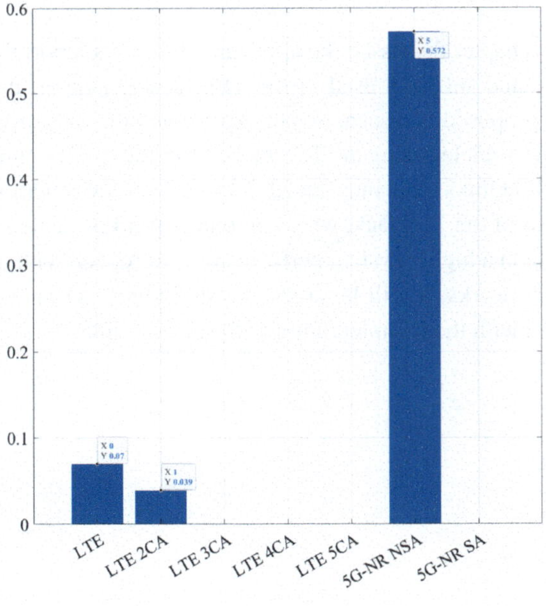

(b)

References

1. T. Bilen and B. Canberk, "Learning-vector-quantization-based topology sustainability for clustered-aanets," *IEEE Network*, vol. 35, no. 4, pp. 120–128, 2021.
2. ACI World, "ACI world airport traffic forecasts 2022-2041," https://store.aci.aero/form/2022-watf-executive-summary/, 2023.
3. T. Bilen, H. Ahmadi, B. Canberk, and T. Q. Duong, "Aeronautical networks for in-flight connectivity: A tutorial of the state-of-the-art and survey of research challenges," *IEEE Access*, vol. 10, pp. 20 053–20 079, 2022.
4. S. Al-Rubaye and A. Tsourdos, "Airport connectivity optimization for 5G ultra-dense networks," *IEEE Transactions on Cognitive Communications and Networking*, vol. 6, no. 3, pp. 980–989, 2020.
5. S. K. Khan, U. Naseem, A. Sattar, N. Waheed, A. Mir, A. Qazi, and M. Ismail, "UAV-aided 5G Network in Suburban, Urban, Dense Urban, and High-rise Urban Environments," in *2020 IEEE 19th International Symposium on Network Computing and Applications (NCA)*, 2020, pp. 1–4.
6. G. Geraci, A. Garcia-Rodriguez, M. M. Azari, A. Lozano, M. Mezzavilla, S. Chatzinotas, Y. Chen, S. Rangan, and M. D. Renzo, "What will the future of UAV cellular communications be? a flight from 5G to 6G," *IEEE Communications Surveys & Tutorials*, vol. 24, no. 3, pp. 1304–1335, 2022.
7. P. Mohapatra, J. Li, and C. Gui, "QoS in mobile a hoc networks," *IEEE Wireless Communications*, vol. 10, no. 3, pp. 44–52, 2003.
8. F. Gianfelici, "Measurement of quality of service (QoS) for peer-to-peer networks," in *IEEE Symposium on Virtual Environments, Human-Computer Interfaces and Measurement Systems, 2005.*, 2005, pp. 6 pp.–.
9. S. Horsmanheimo, L. Tuomimäki, V. Semkin, S. Mehnert, T. Chen, M. Ojennus, and L. Nykänen, "5G communication QoS measurements for smart city UAV services," in *2022 16th European Conference on Antennas and Propagation (EuCAP)*, 2022, pp. 1–5.
10. S. Chan and V. Chan, "Constellation topologies for a space-based information network backbone using optical intersatellite links," in *IEEE MILCOM 2004. Military Communications Conference, 2004.*, vol. 2, 2004, pp. 812–821 Vol. 2.
11. D. Tomic, S. Hofmann, M. Ozger, D. Schupke, and C. Cavdar, "Quality of service aware traffic management for aircraft communications," 2020.
12. Q. Luo and J. Wang, "Multiple qos parameters-based routing for civil aeronautical ad hoc networks," *IEEE Internet of Things Journal*, vol. 4, no. 3, pp. 804–814, 2017.
13. R. Martínez Vidal, R. Martí, C. Sreenan, and J. Borrell, "Measuring QoS in an aeronautical opportunistic network architecture with limited access to a satellite communications backhaul," *Mobile Information Systems*, vol. 2016, pp. 1–12, 01 2016.
14. E. Dinc, M. Vondra, S. Hofmann, D. Schupke, M. Prytz, S. Bovelli, M. Frodigh, J. Zander, and C. Cavdar, "In-flight broadband connectivity: Architectures and business models for high capacity air-to-ground communications," *IEEE Communications Magazine*, vol. 55, no. 9, pp. 142–149, 2017.
15. S. Ayaz, F. Hoffmann, C. Sommer, R. German, and F. Dressler, "Performance evaluation of network mobility handover over future aeronautical data link," in *2010 IEEE Global Telecommunications Conference GLOBECOM 2010*, 2010, pp. 1–6.
16. V. Towhidlou, S. Al-Rubaye, and A. Tsourdos, "LTE handover design for cellular-connected aircraft," in *2022 IEEE/AIAA 41st Digital Avionics Systems Conference (DASC)*, 2022, pp. 1–5.
17. O. S Oubbati, M. Atiquzzaman, P. Lorenz, M. H. Tareque, and M. S. Hossain, "Routing in flying ad hoc networks: Survey, constraints, and future challenge perspectives," *IEEE Access*, vol. 7, pp. 81 057–81 105, 2019.

18. E. Temprado, L. Goratti, A. Albanese, C. Meani, P. Paglierani, O. Segou, E. Kafetzakis, C. Xilouris, and M. R. Spada, "In-flight entertainment and connectivity in the 5G era: the 5G ESSENCE experimental platform," in *2019 European Conference on Networks and Communications (EuCNC)*, 2019, pp. 241–245.
19. P. P. Ray, "A review on 6G for space-air-ground integrated network: Key enablers, open challenges, and future direction," *Journal of King Saud University - Computer and Information Sciences*, vol. 34, no. 9, pp. 6949–6976, 2022.
20. X. Liu, H. Zhang, M. Sheng, W. Li, S. Al-Rubaye, and K. Long, "Ultra dense satellite-enabled 6G networks: Resource optimization and interference management," *China Communications*, pp. 1–14, 2023.
21. X. Li, H. Zhang, H. Zhou, N. Wang, K. Long, S. Al-Rubaye, and G. K. Karagiannidis, "Multi-agent drl for resource allocation and cache design in terrestrial-satellite networks," *IEEE Transactions on Wireless Communications*, vol. 22, no. 8, pp. 5031–5042, 2023.
22. D. Roddy, *Satellite Communications*, ser. McGraw-Hill telecom engineering. McGraw-Hill, 2001.
23. "ARINC 781 SDU, SCM, and FMPA system description, installation, and maintenance manual," 21 Oct 2014. [Online]. Available: https://usermanual.wiki/EMS-Technologies-Canada/FMPA-3703506.pdf
24. G.-S. Ahn, A. Campbell, A. Veres, and L.-H. Sun, "Supporting service differentiation for real-time and best-effort traffic in stateless wireless ad hoc networks (SWAN)," *IEEE Transactions on Mobile Computing*, vol. 1, no. 3, pp. 192–207, 2002.
25. E. Gemikonakli, E. Ever, G. Mapp, and O. Gemikonakli, "Admission control and buffer management of wireless communication systems with mobile stations and integrated voice and data services," *Telecommunication Systems*, vol. 65, no. 4, pp. 663–675, 2017.
26. D. Marinos, C. Aidinis, C. Vassilopoulos, P. Kouros, N. Schmitt, T. Pistner, J. Schalk, and J. Klaue, "Performance analysis of modulation schemes for wireless optical in-cabin aircraft networks," in *2010 International Symposium on Information Technology*, vol. 3, 2010, pp. 1401–1403.
27. S. Ayub, I. Petrunin, A. Tsourdos, S. Al-Rubaye, G. Stapylton, and G. Dent, "In-flight entertainment datalink analysis and simulation," in *2020 AIAA/IEEE 39th Digital Avionics Systems Conference (DASC)*, 2020, pp. 1–10.
28. H. Whitworth, S. Al-Rubaye, A. Tsourdos, J. Jiggins, N. Silverthorn, and K. Thomas, "Aircraft to operations communication analysis and architecture for the future aviation environment," in *2021 IEEE/AIAA 40th Digital Avionics Systems Conference (DASC)*, 2021, pp. 1–8.
29. S. Al-Rubaye, A. Tsourdos, and K. Namuduri, "Advanced air mobility operation and infrastructure for sustainable connected evtol vehicle," *Drones*, vol. 7, no. 5, 2023.
30. S. Mondal, S. Al-Rubaye, and A. Tsourdos, "Handover prediction for aircraft dual connectivity using model predictive control," *IEEE Access*, vol. 9, pp. 44 463–44 475, 2021.
31. M. Ozpolat, S. Al-Rubaye, A. Williamson, and A. Tsourdos, "Integration of unmanned aerial vehicles and LTE: A scenario-dependent analysis," in *2022 International Conference on Connected Vehicle and Expo (ICCVE)*, 2022, pp. 1–6.
32. A. Warrier, S. Al-Rubaye, D. Panagiotakopoulos, G. Inalhan, and A. Tsourdos, "Interference mitigation for 5G-connected UAV using deep Q-learning framework," in *2022 IEEE/AIAA 41st Digital Avionics Systems Conference (DASC)*, 2022, pp. 1–8.
33. A. J. Ali, M. Khalily, A. Sattarzadeh, A. Massoud, M. O. Hasna, T. Khattab, O. Yurduseven, and R. Tafazolli, "Power budgeting of LEO satellites: An electrical power system design for 5G missions," *IEEE Access*, vol. 9, pp. 113 258–113 269, 2021.
34. B. A. Homssi, K. Dakic, K. Wang, T. Alpcan, B. Allen, R. Boyce, S. Kandeepan, A. Al-Hourani, and W. Saad, "Artificial intelligence techniques for next-generation mega satellite networks," 2023.

35. Electronic Communications Committee (ECC) Reports, "The use of the frequency bands 27.5-30.0 GHz and 17.3-20.2 GHz by satellite networks," Sept 2010.
36. K. W. Fischer, M. R. Witiw, J. A. Baars, and T. R. Oke, "Atmospheric laser communication: New challenges for applied meteorology," *Bulletin of the American Meteorological Society*, vol. 85, no. 5, pp. 725 – 732, 2004.
37. J. Zhang, T. Chen, S. Zhong, J. Wang, W. Zhang, X. Zuo, R. G. Maunder, and L. Hanzo, "Aeronautical *ad hoc* networking for the internet-above-the-clouds," *Proceedings of the IEEE*, vol. 107, no. 5, pp. 868–911, 2019.
38. H. Dahmouni, A. Girard, M. Ouzineb, and B. Sanso, "The impact of jitter on traffic flow optimization in communication networks," *IEEE Transactions on Network and Service Management*, vol. 9, no. 3, pp. 279–292, 2012.

NOMA-Enabled RIS-Assisted SATINs Under Multi-objective Optimization: A Deep Reinforcement Learning Approach

Kefeng Guo, Min Wu, Xingwang Li, Ali Nauman, Changqing Li, and Mingfu Zhu

Abstract

Satellite-aerial-terrestrial integrated networks (SATINs) with reconfigurable intelligent surface (RIS) and non-orthogonal multiple access (NOMA) are promising for next-generation communications due to their ability to intelligently reconfigure transmission

K. Guo · M. Wu · C. Li
School of Space Information, Space Engineering University, Beijing 101400, China
e-mail: guokefeng.cool@163.com

M. Wu
e-mail: 1800022837@pku.edu.cn

C. Li
e-mail: lcqqcl5577@sohu.com

X. Li (✉)
School of Physical and Electronics Engineering, Henan Polytechnic University, Jiaozuo 454003, China
e-mail: lixingwang@hpu.edu.cn

A. Nauman
School of Computer Science and Engineering, Engineering, Yeungnam University, Gyeongsan 38541, South Korea
e-mail: anauman@ynu.ac.kr

M. Zhu
College of Computer Science and Technology, Henan Polytechnic University, Jiaozuo 454003, China
e-mail: zhumingfu@zzu.edu.cn

© The Author(s), under exclusive license to Springer Nature Switzerland AG 2025
M. A. Jamshed and A. Nauman (eds.), *Integrated Terrestrial and Non-Terrestrial Networks*, Synthesis Lectures on Communications,
https://doi.org/10.1007/978-3-031-63705-6_7

environments and improve spectral efficiency, but face the tricky challenge of multi-objective optimization of network performance in dynamic communication environments. To better describe the model, a multi-vector optimization problem (MVO) is proposed to improve system reachability under many constraints, considering UAV energy, transmit beamforming, and RIS phase-shift. Real-time interaction with the environment is achieved with DRL algorithms, using multi-vector deep deterministic policy gradient (MV-DDPG) to find a suboptimal solution. Experimental results show an effective adjustment of the optimal update policy when considering three objectives simultaneously and different weighting settings.

Keywords
Deep reinforcement learning • Reconfigurable intelligent surface • Satellite-aerial-terrestrial integrated networks (SATINs) • Reconfigurable intelligent surface • Non-orthogonal multiple access • Multi-vector DDPG

1 Introduction

According to business reports, the total number of ground terminal connections is predicted to reach 27.1 billion globally by 2025 [1]. This increasing demand for widespread access to the Internet of Things (IoT) has led to a demand for next-generation networks that offer global coverage, on-demand access to services, and robust security and reliability. To address these needs, there is ongoing research into developing an integrated framework for satellite-air-ground networks SATINs. This framework aims to provide comprehensive connectivity and support for IoT applications, catering to the evolving requirements of modern communication networks [2]. Specifically, the primary objective of SATINs is to provide comprehensive and effective collaborative processing of events. This can be achieved by utilizing multidimensional information and working in conjunction with various modules to enhance event processing capabilities. The system is an organic integration of different networks, such as satellite communications (SatComs), including deep space networks, space vehicles, and terrestrial-based wired and wireless facilities with multiple functions, such as communication, reconnaissance, navigation, and meteorology. These networks connect users, vehicles, and various communication platforms on the ground, in the air, and deep space via inter-satellite and satellite-terrestrial links. Interplanetary links can also be established between the ground and satellites for data exchange based on application requirements. These networks can be either an on-demand integration of existing satellite systems or the result of integrated system design according to demand, featuring multi-functional integration, dynamic and variable compositions, complex operation status, and integrated information exchange and processing. Besides, SATINs can yield efficient and effective collaboration, while providing a broad range of applications in communication, reconnaissance, navigation, and meteorology. In summary, SATINs offer significant advantages in event processing through the integration of multidimensional information, diverse working modules, and different networks.

However, at the same time, it is also the case that the transmission quality of SATINs can be easily affected by the transmission environment, as there is a high probability that there are many obstacles in the signal transmission path in the case of long transmission distances. To ensure service reliability and consistency of SATIN communications in the face of high heterogeneity, it is essential to carefully analyze the characteristics of the various wireless channels involved. In addition, effective mitigation strategies must be developed to address the challenges posed by environmental factors [3].

1.1 For the RIS-Assisted SATINs

In recent years, the rapid development of information technology tools has led to an explosive growth in the volume of services in mobile communication networks. With more and more people rely on their mobile devices for various tasks, such as communication, entertainment, and accessing information, the demand for mobile services has skyrocketed. However, this growth in mobile communication networks has also come with its own set of challenges, particularly in terms of energy consumption. As the size of networks has expanded and the number of base stations has increased, the energy consumption of these networks has also significantly risen. Additionally, the diversity of base stations and the randomness of equipment siting have further contributed to the overall energy demands of the network. One of the key issues that has emerged as a result of this increasing energy consumption is power supply challenges for base stations. These challenges arise from the fact that the energy demands of the networks have outpaced the capabilities of existing power supply systems. This has created a need for more sustainable and easily deployable communication networks that can effectively meet the increasing energy demands. The global concern for developing such sustainable and easily deployable communication networks has become a pressing issue. This concern stems from the recognition that the current energy consumption patterns in mobile communication networks are not sustainable in the long run.

To address these challenges, researchers and industry experts have been working on various solutions. These solutions range from the development of energy-efficient network equipment, such as novel antenna design schemes. Additionally, efforts are being made to optimize network infrastructure and operation to minimize unnecessary energy consumption. The advancement in communication technology is accompanied by the rapid development of metamaterials, which have shown great potential in improving wireless signal propagation. One such success story is the reconfigurable intelligent surface (RIS), which enable efficient control of wireless signal propagation by regulating reflection and refraction. By adjusting the phase and amplitude of the RIS elements, it is possible to effectively amplify the desired signal and mitigate interfering signals, thereby improving network coverage and spectral efficiency. This capability enables the creation of so-called "smart radio environments" where advanced wireless communication systems can be developed to meet the growing the communication needs of modern society [4]. This presents a promising solution for integrating into current wireless communication networks due to its elimination of the need for RF chains and high-power components. This advantage translates into improved energy efficiency and cost-effectiveness compared to traditional active relaying technologies. By incorporating RIS into existing networks, we can easily achieve multiple goals, such as enhancing energy efficiency and reducing inter-user interference.

The advantages offered by RIS technology have led to extensive research on RIS-assisted SATINs. One application involves deploying RIS at a fixed mounting position, such as a building, to expand coverage or transmit signals to areas with signal blockage. Additionally, the use of UAVs as mobile platforms for RIS (UAV-RIS) presents exciting possibilities. The served users in UAV-RIS-aided networks can benefit from panoramic reflections, which significantly enhance line-of-sight transmission links and extend coverage areas. This ultimately leads to increased network throughput. By combining fixed deployment strategy and mobile RIS solutions, we can effectively enhance IoT communications in various settings, including vehicles, indoor environments, maritime areas, factories, and dense urban areas. Figure 1 illustrates the potential of this approach.

Extensive research has been conducted to analyze the performance of RIS-aided networks. For instance, researchers in [5] investigated the UAV-RIS scheme in a covert commu-

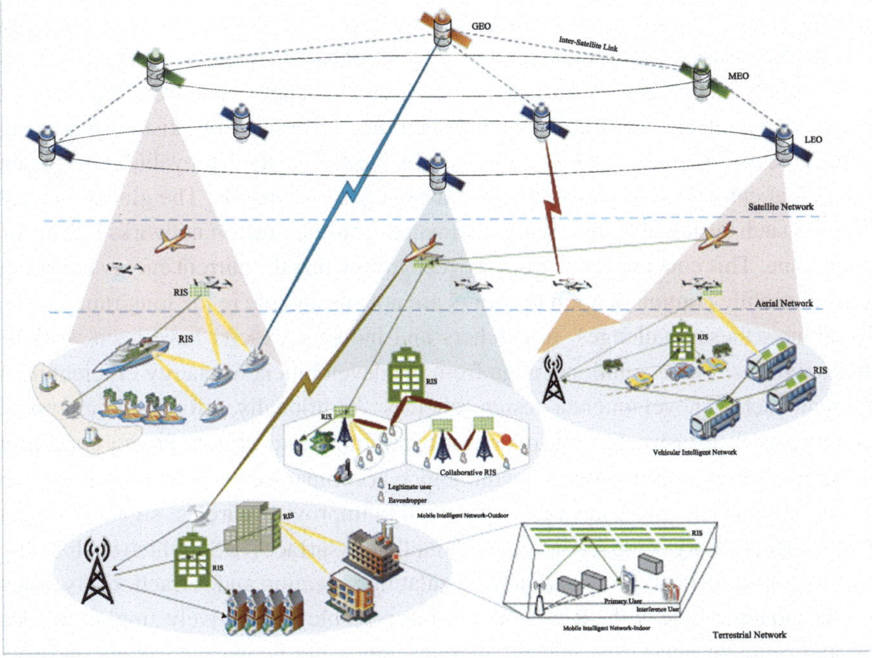

Fig. 1 The framework of RIS-aided SATINs

nication scenario to maximize the covert transmission rate. Another study by [6] considered RIS-aided air-to-ground (A2G) systems in two different communication scenarios. The first scenario involves mounting the RIS on the side of the UAV to establish a line-of-sight link between the transmitter and receiver, while the second focuses on enhancing the signal of the target user and suppressing eavesdropping users by intelligently setting the phase shift of the RIS. The RIS is employed to manipulate the signal propagation environment, thereby improving the signal quality for the desired user and mitigating the impact of unwanted users. In [7], the authors examined the outage probability and system average sum rate with the aid of multiple RIS. Besides, the issue of ensuring secure communication in satellite-ground relay networks with multiple eavesdroppers was addressed in [8] with the help of RIS. And, the authors also focused on tackling the challenges associated with maintaining secrecy and preventing unauthorized access in such networks. These studies highlight the advancements in analyzing and optimizing the performance of various scenarios, contributing to the development and implementation of RIS technology.

1.2 An Overview of RIS-Assisted SATINs

The integration of multiple networks in SATIN makes it inherently more complex than existing classical communication networks such as terrestrial cellular networks. This complexity arises from the need to optimize different RIS deployment strategies and network scheduling algorithms to accommodate the diverse requirements of various applications. Given the dynamic nature of SATIN, the process of optimizing RIS-assisted systems becomes even more intricate. Synergistic algorithms are employed to maximize the benefits of different RIS deployment strategies, while network scheduling algorithms must consider the diverse priorities of various applications. These challenges add further intricacy to the optimization of RIS-assisted SATIN systems.

1.2.1 Optimization Aspects of RIS-Assisted SATINs

Spectrum Management: To satisfy the requirements of different user connections and high speeds, RIS-assisted SATINs need to utilize multiple frequency bands working in tandem [9, 10]. However, satellites have limitations in terms of frequency resources, transmit and receive power management, and signal processing capabilities due to size, weight, and energy constraints. This presents a challenge for large-scale computational capacity when optimizing the performance of RIS-assisted SATINs, especially with the future development of low earth orbit constellations facing limited spectrum resources and trusted interconnection of everything due to rapid and large-scale deployment of commercial spaceflight. Another challenge in optimizing RIS-assisted SATINs is the consideration of different communication processes resulting from using the same wireless channel or adjacent channels. The inherent heterogeneity and high mobility of RIS-assisted SATINs introduce complexities that require more efficient spectrum management techniques. The authors in [11] emphasized that these complexities needed to be fully considered when optimizing the performance of RIS-assisted SATINs.

Energy Management: RIS can be installed in various locations such as satellites, airborne platforms, terrestrial buildings, or indoors to extend communications coverage. However, unlike terrestrial installations that can be easily connected to an energy source, airborne and satellite infrastructures, as well as certain IoT edge devices, face challenges in terms of power supply. They cannot be easily connected to power stations, making the energy utilization of a fully charged battery crucial for their uptime during communication missions. Additionally, the space communications environment presents further challenges in terms of energy consumption. Influencing factors such as radiation intensity and temperature variations in space can impact energy efficiency. In the case of airborne platforms carrying RIS, power consumption is primarily allocated for flight and to support maneuverability. Therefore, efficient energy management becomes essential to balance the power requirements for both auxiliary communications and flight operations.

Routing and Handover Management: RIS-assisted SATIN systems are characterized by their multi-layer heterogeneous network architecture, which enables the creation of a flexible, resilient, and high-capacity integrated information network [12]. However, certain scenarios like LEO high-speed motion introduce significant challenges to the design of routing strategies. The fast movement of satellite beams over the ground at several kilometers per second causes Doppler frequency shifts and dynamic changes in network connectivity. This necessitates careful consideration of factors such as packet loss rate and end-to-end delay, particularly for long transmission paths. Moreover, the high mobility of satellites and UAVs introduces uncertainty in the location of mobile edge devices, making it challenging to ensure seamless signal transmission. Consequently, frequent switching between routes and links becomes necessary. Efficient routing strategies that can adapt to rapidly changing topologies and support high mobility are crucial in addressing these challenges. Furthermore, the design of RIS-assisted SATINs must take into account the unique characteristics of each installation model and the impact of environmental factors on signal transmission. These factors include variations in terrain, weather conditions, and interference sources. To maximize the performance of RIS-assisted SATINs, it is essential to optimize signal transmission considering these environmental influences.

1.2.2 Challenges to RIS-Assisted SATINs

The performance optimisation of RIS-assisted SATINs is summarised in the following practical difficulties:

Interference with multiple devices: RIS-assisted SATINs must also address co-frequency interference between satellite communication systems and terrestrial cellular systems. This interference arises in both the uplink and downlink, due to multibeam satellite antenna beam co-frequency multiplexing and co-frequency multiplexing techniques used in cellular cells under the coverage of the satellite communication system. With the introduction of RIS, the passive beam formed by the reflected elements and the active beam direction of the transmitting segment become key elements that can significantly impact network performance. Efficient interference management techniques are essential to minimize the impact of interference on the overall network performance. This includes strategies such as dynamic resource allocation, interference cancellation, and beamforming optimization.

Computational Complexity: Optimizing the transmit active beamforming matrix, RIS phase-shift matrix, and trajectory strategy in RIS-assisted SATINs involve dealing with a high degree of computational complexity. This complexity arises from the large number of RIS scattering elements and the mobility of UAVs/satellites, which increases the dimensionality of the optimization problem. The challenges become even more pronounced in scenarios involving multi-user or multi-cell communication, where various factors, such as RIS positioning strategy, user scheduling, power allocation, and band selection/switching, must be taken into account. To address these complexities, efficient algorithms that can provide smarter convex approximation and decomposition designs are required. Such algorithms must be capable of handling the intricacies of optimizing RIS-assisted SATINs in dynamic and diverse communication scenarios. These algorithms should be able to adapt to changes in network topologies and efficiently allocate resources to ensure the seamless delivery of services. Additionally, they must be scalable to support the rapidly growing demands of modern communication networks.

Imprecise Channel Modeling: The presence of a large number of passive components in RIS, without signal processing capability, pose challenges for traditional channel estimation methods in RIS-assisted SATINs [13]. RIS, when integrated into millimeter-wave or (sub)terahertz communication systems, plays a crucial role in facilitating line-of-sight connectivity. These frequency bands are known for their high directionality and sparsity in the wireless channel, with only a limited number of resolvable paths. As a result, the channel coefficient matrix becomes rank-deficient, leading to inherent channel sparsity. By leveraging this channel sparsity, RIS can enhance the performance of communication systems in these frequency bands. The RIS elements are strategically placed to manipulate the phase of the incident signals, enabling beam steering techniques to establish LoS links between the transmitter and receiver. This helps overcome the limitations imposed by obstacles and multipath fading, thereby improving the overall system capacity and reliability. To address this, compressed sensing techniques are employed in RIS-assisted systems for more accurate channel estimation. These techniques include atomic-paradigm minimization, basis tracking, approximate message passing, and hybrid-paradigm minimization [14, 15].

1.3 RIS-Assisted SATINs: A Deep Reinforcement Learning Scheme

The discrete and randomly distributed nature of the large number of scattering elements in RIS presents significant challenges for evaluating and optimizing RIS-assisted radio systems. These challenges extend beyond traditional issues related to network security, resource allocation, and channel fading. For example, in RIS-assisted SATIN, the movement of UAVs can cause beam alignment angle misalignment problems, resulting in degraded performance of beamforming techniques. Consequently, designing a scheme for dynamic beamforming alignment has become a focus of research, which aims to improve the reliability of RIS-assisted SATIN through continuous interaction with the environment and real-time correction of beam misalignment. To address this challenge, a significant body of literature has focused on approximating the dynamic beamforming alignment problem as a convex optimization problem using methods such as alternating optimization (AO) and semi-definite relaxation (SDR). However, due to the structural design of the RIS, the dimensionality of the phase-change data increases, and conventional numerical analysis modeling based on these methods becomes computationally time-consuming. This is primarily due to the need for high-resolution control over the phase shifts of the numerous RIS elements, which results in a large number of variables and complex optimization formulations.

The phase superposition of multiple elements in RIS presents a high-dimensional joint beamforming design problem, which can be formulated as a mixed-integer high-dimensional nonlinear programming problem. Conventional algorithms often face challenges in solving this complex problem effectively. In dynamic RIS configuration scenarios, where the reflection coefficients are assumed to be reconfigured multiple times during a transmission instance, determining the capacity region becomes prohibitively complex due to the exponential growth in the size of the set of all possible combinations of reflection coefficients with each reconfiguration. In large-scale RIS-assisted SATINs, where users are constantly moving within the network, dynamically optimizing the location of RIS (e.g., by mounting RISs on UAVs) and selecting the appropriate user access scheme and RIS configuration holds promise for meeting the time-variant quality of service requirements of the users, especially in the context of real-time IoT or Internet of Vehicles (IoV) applications [16–18]. However, the highly dynamic and complex channel environment of future SATINs poses significant computational challenges for conventional methods, leading to long optimization times.

In RIS-assisted SATIN systems, various applications require different priority designs in network scheduling algorithms, which can complicate network optimization. To address these challenges, machine learning (ML) based on a long-term performance metric can be a suitable solution [19, 20]. Deep learning (DL) and reinforcement learning (RL) are subdivisions of machine learning, with deep learning requiring massive datasets for training, while reinforcement learning constantly adjusts network parameters based on trial and error and external interactions. Deep reinforcement learning (DRL), which combines the characteristics of deep learning and reinforcement learning, has been proven to be a revolutionary breakthrough with a higher-level understanding of the visual world [21]. DRL has emerged as a powerful approach for solving high-dimensional problems by learning low-dimensional feature representations. Its combination with neural networks allows for efficient function approximation, enabling the applications of DRL in various domains [22, 23]. In wireless communication systems, DRL can effectively address the curse of dimensionality problem by learning feature covariates in the wireless channel through representations [24, 25]. This is different from traditional non-parametric approaches that rely on tabular formats. DRL is particularly suitable for large-scale multiple-input multiple-output (MIMO) systems, such as RIS-assisted systems with a large number of array elements or SATIN. These systems involve optimization problems with an extremely large number of dimensions, making them intractable using conventional methods [26, 27]. By leveraging DRL, the optimization problem can be solved efficiently without the need for exhaustive sample libraries or explicit models. Offline prediction can be used to reduce complexity and computation time. One advantage of DRL over deep learning (DL)-based methods is that it does not require a model and can learn and build knowledge about the radio channel in real-time, even in the absence of channel models or mobility patterns [28, 29]. This allows wireless communication systems to efficiently adapt and find solutions to complex optimization problems by observing environmental rewards. DRL provides a flexible and adaptive algorithm design approach for wireless communication systems. Many previous works have applied DRL in wireless communication systems, including obtaining hybrid beamforming matrices at the transmitter in [30], joint design of beamforming, power control, and interference coordination to maximize SINR in [31], cell vectorization problem for optimal beamforming matrix selection in [32], trajectory design of multiple UAVs based on user mobility information in [33], and optimization of passive phase shift of RIS each element to maximize downlink received signal-to-noise ratio in [34].

1.3.1 Future Directions

In this section, we will continue to look forward to the development direction of DRL-enabled RIS-assisted SATINs [35, 36].

Computation Efficiency Tradeoff: The effectiveness of DRL techniques is closely tied to continuous interaction with the wireless communication environment, allowing for the acquisition of feedback, such as rewards or penalties. However, this ongoing interaction can result in heightened computational demands and increased resource consumption. Therefore, it's imperative to thoroughly explore the algorithmic framework in order to enhance the computational efficiency of DRL methods. This exploration should encompass various factors, including the design of neural network architecture, training strategies for network optimization, and the performance of computer hardware. When studying the impact of these factors, it is important to consider the trade-offs involved. For instance, while increasing the number of layers in a neural network may enhance the accuracy of estimating the actual optimization objective, it also escalates computational requirements. As a result, a comprehensive investigation is essential to strike a balance between computational efficiency and estimation accuracy, taking into account the specific requirements and constraints of the wireless communication system. To mitigate the computational complexity, offline learning methods can be employed. Offline learning reduces the amount of computation required by utilizing pre-collected data for training. On the other hand, online learning methods enable the DRL architecture to adapt to changes in the environment. This adaptability comes at the expense of increased computational requirements, as the network continuously learns and updates its policies based on real-time interactions with the environment.

Framework Construction: One of the primary challenges in DRL-based schemes is establishing the appropriate DRL architecture for different communication environments. The introduction of RIS in SATINs has transformed the network into a multi-dimensional system, significantly increasing its complexity compared to traditional ground-based networks. In the DRL-enabled optimization framework, the architecture involves learning inner-loop neural networks and interacting with the outer-loop environment through rewards or punishments. This adds another layer of complexity to the design process. Finding the right balance between complexity and efficiency is key to developing effective DRL architectures for RIS-assisted communication systems. This requires thoughtful consideration of the network's specific characteristics and the trade-offs involved in the learning process. By properly designing the input and output mechanisms, we can enhance the efficiency of learning while ensuring the system behaves optimally based on the desired rewards.

Training and Performance Evaluation: Coordinating multiple RISs in SATIN to optimize network performance often involves training neural networks to determine the optimal policy. Constructing these models requires a significant amount of real communication channel data. However, compared to other DRL applications like computer vision, SATIN faces unique challenges due to constraints such as unstable energy sources and limited storage capacity in wireless communication systems [37]. These challenges are compounded by the stochastic nature of channels and data arrivals in wireless environments, making it extremely difficult to acquire sufficient energy for communication. Consequently, there is a scarcity of extensive reference data pools in SATIN's real communication data. As a result, current research heavily relies on simulated communication data to compensate for this limitation. However, using simulated data may oversimplify the complexity of the real system and overlook hidden patterns and nuances that exist in actual wireless environments. To address these challenges, it is crucial to explore techniques that can mitigate the limitations imposed by energy constraints and limited data availability. This could include developing advanced energy acquisition techniques that improve the collection of real communication channel data while considering resource limitations. Additionally, efforts should be made to enhance the fidelity of simulated data by incorporating more realistic models and capturing the intricacies of wireless communication systems.

2 Introduction to RIS-Assisted Communication: A NOMA-Enabled Approach

The next generation of RIS-assisted SATINs aims to not only enhance the spectral efficiency but also support massive device connectivity. Simultaneously, it poses challenges in simplifying system design and signaling processes, which can be a significant challenge for existing orthogonal multiple access technologies. To address these challenges, the proposal of non-orthogonal multiple access (NOMA) technology has emerged. NOMA technology introduces a paradigm shift by allowing multiple users to share power resources instead of exclusive resource allocation to a single user in the power domain. At the receiving end, interference cancellation techniques are employed to distinguish and decode signals from different users. This approach enables efficient utilization of available resources and enhances the capacity of the system.

By incorporating NOMA technology into RIS-assisted SATINs, it is anticipated that the next generation of systems will achieve higher spectral efficiency, accommodate a large number of connected devices, simplify system design, and streamline signaling processes. However, the successful implementation of NOMA in RIS-assisted SATINs require further research and development to overcome technical challenges and ensure optimal performance.

2.1 Overview of the Basic Concepts of NOMA Technology

The NOMA-based scheme is to improve spectral efficiency by enabling multiple users to share the same radio frequency resources in the same time and frequency domains. This section focuses on NOMA access strategies under power domain multiplexing. This subsection will focus on these two technologies and illustrate the advantages and disadvantages of applying NOMA in RIS-assisted SATINs.

2.1.1 Key Techniques of NOMA Technology

Here, we briefly discuss two key technologies for realising NOMA, including superposition coding (SC) and successive interference cancellation (SIC).

- *SC technique*: This technique belongs to the family of information encoding techniques and was initially introduced by Cover in 1972 [38]. The primary objective of SC is to utilize the same bandwidth resource to transmit information to multiple receivers through a single reflector, allowing for overlapping communication. Since its inception, this technique has found applications in various channels, including interference, relay, and wiretap channels. Its effectiveness was initially evaluated through experimental studies conducted in software-defined networks (SDNs) [39]. The core principle behind SC involves combining the messages of multiple users linearly in the power domain before transmission. This enables the share of the same radio resources among multiple users.

- *SIC technique*: The successive interference cancellation (SIC) technique involves sorting signals based on the strength of the channel state information, allowing the receiver to decode the signal with the highest quality first, subtract it, and then proceed to decode the signals from weaker users. The complexity of SIC is directly influenced by the density of users sharing the same frequency/time resource. As the number of devices on the same frequency/time resource increases, the processing complexity of SIC also rises. This underscores the importance of managing signal power levels to ensure effective operation of SIC within NOMA-based communication systems [40].

2.2 RIS-Assisted Communication Exploiting NOMA Schemes

Due to the vast potential benefits offered by RIS-assisted communications, researchers have started exploring the combination of RIS with multiple access schemes. This innovative approach introduces new spatial freedom, allowing intelligently coordinated interference to serve multiple users simultaneously in the time, frequency, and power domains. The above analysis highlights the potential of NOMA as an access scheme for RIS-assisted SATIN. NOMA has the ability to mitigate the effects of multipath propagation and shadow fading, making it a promising solution for wireless communication systems [41, 42]. In particular, recent works such as [43, 44] have investigated the performance of NOMA networks in the presence of residual transceiver hardware deficiencies, comprehensively exploring outage probability (OP) and ergodic capacity(EC) in the presence of both cooperative or non-cooperative user networks. These studies mentioned above provide valuable insights into the practical implementation of NOMA in real-world communication systems. However, further research is needed to fully understand the benefits and limitations of NOMA as an access scheme for RIS-assisted SATIN, taking into account the specific requirements and constraints of different wireless communication scenarios.

Figure 2 depicts various RIS-assisted networks utilizing NOMA access, emphasizing two key RIS configurations: static and dynamic. In the static RIS configuration, the RIS's reflection coefficient is adjusted once during a transmission mission, and users are divided based on the static NOMA policy in the power, time, and frequency domains. Conversely, the dynamic RIS configuration involves reconfiguring the transmission coefficients across multiple time slots within a single transmission task. However, as the dimension of the RIS reflection elements increases, so does its computational complexity, posing new challenges for researchers. The integration of RIS with multiple access schemes represents a promising area for future research, offering the potential to revolutionize wireless communication by simultaneously serving multiple users, effectively managing interference, and maximizing spectral efficiency.

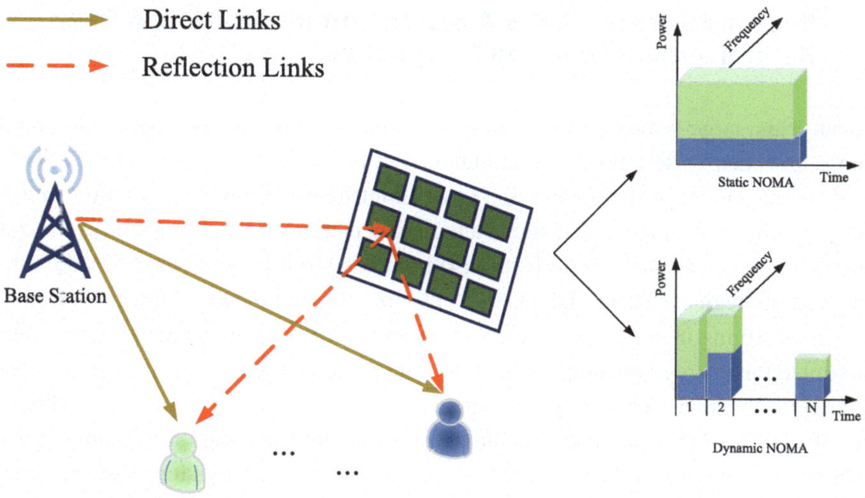

Fig. 2 NOMA-empowered scheme in RIS-assisted networks for static and dynamic RIS configurations

2.3 Benefits of RIS-Assisted Communication in NOMA-Based Scenarios

In addition, the integration of dynamically configurable RISs into NOMA-based networks offers several advantages:

- *1*: RISs are simple to integrate into existing NOMA networks with significant performance improvements. By providing additional signal diversity to the existing wireless channel, RISs can enhance the received signal quality and help mitigate interference among users. This is achieved without requiring additional time slots or energy, as RISs, simply reflect the incoming signals with a specific phase shift.
- *2*: NOMA networks with RISs offer greater design flexibility and the ability to transition from "channel condition-based NOMA" to "QoS-based NOMA". As a technology that can dynamically change the channel, RIS can adjust the reflection coefficient and deployment location to improve or degrade the channel quality for individual users, resulting in intelligent NOMA-based communication scenarios design.
- *3*: RIS can reduce the limitations of a multi-antenna NOMA-empowered system. This is because in a conventional MIMO-NOMA transmitter network, limits on the number of antennas at the transmitter and receiver may need to be met. However, the additional passive array gain provided by RIS can alleviate these limitations.

In summary, the incorporation of RIS into NOMA-based networks provides new opportunities and challenges for wireless communication performance in multi-user environments.

3 Recent Advances in RIS-Assisted Communication: A Deep Reinforcement Learning Perspective

In recent years, large-scale data processing technology, represented by deep reinforcement learning, has been widely used in communication networks, such as resource allocation, resource management [45], transmission power control, user demand prediction, and so on [46, 47]. Aiming at the problem of signal anomaly detection and multi-criteria optimization. Taking this as an example, this project in [48] intended to use the method based on depth-determined strategic gradient (DDPG) to optimize the sub-transmission power and RIS active beamforming in cognitive wireless communication. In [49], as the research object is to maximize the energy consumption of UAVs through the comprehensive optimization of UAV motion, RIS phase change, power allocation, and decoding sequence. In addition, the phase offset of the UAV can reach the maximum value by adjusting the 3D moving path of the RIS and the position of the terminal to the ground. Therefore, dynamic distance learning is an effective method to process information by using external information.

4 Beamforming Optimization and Trajectory Design in NOMA-Based Multi-objective RIS-Assisted SATINs: A Deep Reinforcement Learning Method

Combined with the actual needs of the development of the new generation mobile communication technology in our country, this topic carries out the downlink SATINS research of non-orthogonal multiple access technology based on RIS. Its purpose is to maximize the achievable data transmission rate on the premise of reducing the energy consumption of UAVs. In order to solve this problem, this paper uses the DDPG method to track UAVs in real-time. On this basis, a new reward function model based on multi-dimensions is proposed. Therefore, the neural network has good fitting performance and can effectively overcome non-linear optimisation problems. In this method, the optimal solution is obtained by adaptively searching the weighting coefficients of each aircraft. At the same time, this method is also suitable for solving multiple objective functions. The proposed approach takes into account the performance convergence issues that may arise due to different learning rates of neural networks when encountering local extremes. This consideration ensures that the algorithm can effectively navigate towards globally optimal solutions while avoiding getting stuck in suboptimal local solutions. In addition, the use of a cooperative optimal track for multi-UAV joint navigation can effectively improve the complex communication capability of RIS to satellite navigation.

4.1 System Model and Problem Formulation

The architecture patterns recommended in this study are shown in Fig. 3. The scheme is a downlink SATIN communication mode based on RIS, which is composed of satellite, airborne RIS (each UAV has one RIS), and multiple vehicle equipment. The satellite uses a single omni-directional antenna for a total of K of single-antenna ground users (GUs) transmission services. In order to solve this problem, this project will give full play to the maneuverability of UAV and actively reconstruct the receiver by using RIS technology, so as to realize the continuous virtual LoS transmission channel to the satellite-workshop. So, we attempt to install a RIS on both sides of the UAV plane, and each RIS contains a reflective unit of M. This can improve the transmission channel between satellite and GU, and improve the quality of wireless communication. Give full play to the respective advantages of UAV and RIS, improve the service quality of the SATIN system, and improve network coverage.

The communication in this system can be divided into two forms, direct transmission form satellite-to-GUs, and reflective signal transmission mode. In the reflective link, the signal is transmitted from the satellite to the reflective RIS on the UAV, which then reflects it to the GUs. It should be noted that in this article, UAV is merely a mobile platform for aerial-mounted RIS and does not participate in the communication itself. To make the expression clearer about the relevant channel, we define the channel model vectors satellite-to-k-th GU, RIS-to-k-th GU, satellite-to-RIS as $\mathbf{G}_{S,K}^{H} \in \mathbb{C}^{1\times M}$, $\mathbf{h}_{R,K}^{H} \in \mathbb{C}^{1\times M}$, and

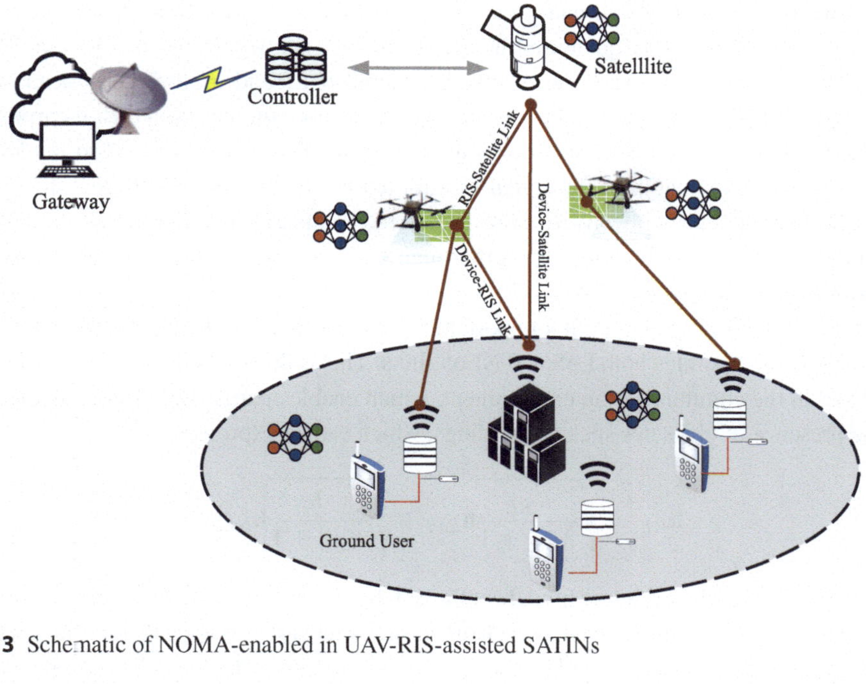

Fig. 3 Schematic of NOMA-enabled in UAV-RIS-assisted SATINs

$\mathbf{G}_{SR} \in \mathbb{C}^{1\times 1}$, respectively. The diagonal phase shift matrix of the i-th RIS is expressed as $\mathbf{\Theta}_i = diag\left(e^{j\theta_i^1}, e^{j\theta_i^2}, ..., e^{j\theta_i^M}\right)$, where $\theta_i^m \in [0, 2\pi)$ denotes the phase shift of the m-th reflecting element. To optimize the performance of the system, we need to jointly optimize the RIS passive beamforming and the satellite active transmission beamforming using a DRL-enabled framework. This optimization needs to take into account the UAV trajectory. This project plans to use ideal CSI to effectively compensate large-scale attenuation, and combine the feedback from the receiver to the receiver, so as to achieve the purpose of suppressing large-scale attenuation. This scheme has been verified in the DVB-S2 standard [50]. Additionally, the perfect CSI can be shared during transmission through feedback and training from the terrestrial GUs.

4.2 Description of Channel Model

Starting from the downlink transmission mode, this project intends to study the coupling relationship between satellite and RIS various elements, including free space loss, rainfall attenuation and beam gain, and comprehensively analyze the effects of various elements on the signal transmission quality of the whole system. Here, it can be shown as

$$\mathbf{G}_\mathbf{s} = \sqrt{G_{s,t}G_{s,p}C_f} \odot \xi^{-\frac{1}{2}} \odot \mathbf{b}^{\frac{1}{2}} \odot e^{-j\phi}, \tag{1}$$

where $G_{s,t}$ denotes the satellite transmitting antenna gain, $G_{s,p}$ stands for the satellite parabolic antenna gain. Additionally, C_f represents the free space loss. Furthermore, the effect of rainfall decay is quantified as ξ_k, specifically satisfying the distribution about $\ln\left(\xi_k^{\text{dB}}\right) \sim \mathcal{CN}\left(\mu_k, \sigma_k^2\right)$. And, the parameters included μ_k and σ_k represent mean value and variance [51]. Here, the vector $\mathbf{b}$ represents the beam gain matrix, which incorporates the far-field characteristics of satellite communication. This matrix is used to model the directionality and strength of the signal transmitted by the satellite in different directions [52]. These parameters collectively contribute to the comprehensive characterization of the downlink channel model, facilitating a thorough understanding of the signal transmission environment in RIS-assisted SATINs.

In the Rician-fading model, it's important to be clear that the channel between the RIS and the GU incorporates both LoS and NLoS links. This is due to the elevated flying height of UAVs in the communication environment, which enables partial LoS transmission even in the presence of obstacles such as buildings, which can be expressed as

$$\mathbf{h}_{RE} = \sqrt{\frac{K_{RE}}{K_{RE}+1}}\mathbf{h}_{LoS} + \sqrt{\frac{1}{K_{RE}+1}}\mathbf{h}_{NLoS}, \tag{2}$$

where K_{RE} is the Rician-fading model coefficient, $\mathbf{h}_{LoS}$ and $\mathbf{h}_{NLoS}$ are the LoS-based channel parameter part and the NLoS-based channel parameter part, respectively [53]. In concrete terms, $\mathbf{h}_{LoS}$ is mainly determined by the position of the UAV, which can be formulated as

$$\mathbf{h}_{LOS} = \sqrt{G_{e,v} C_{e,v}} \text{vec} \left(\mathbf{A} \left(\theta_r, \varphi_r \right) \right) \tag{3}$$

where $G_{e,v}$ is the receive antenna gain, $C_{e,v}$ denotes the path loss between RIS and GUs, and $\mathbf{A}(\theta_r, \varphi_r)$ is a key determinant of transmission performance about the RIS-GUs channel matrix. These parameters play a crucial role in characterizing the spatial orientation and directional properties of the communication links within the RIS-assisted SATINs. The position of the UAV will directly determine the signal transmission angle and the arrival angle of the signal, so it is very critical to optimize the trajectory of the UAV in real time and clarify its position, and thus the proposed algorithmic framework can be dynamically adjusted according to the changes in the communication environment, thus improving the overall performance of the system.

4.2.1 Signal Transmission Model for NOMA Scheme

Conventional OMA schemes face difficulties in balancing spectrum utilization and mass access requirements. NOMA technology offers a solution by allocating the same resources to multiple users through power differentiation, leading to improved spectrum efficiency for future terrestrial vehicular networks. In the NOMA-based downlink system, decoding priority is given to vehicular users with poor channel quality. Specifically, transmit power is distributed over vehicles with poor channel quality rather than over vehicles with good channel state quality [54]. To further simplify the equation below, the composite channel coefficients for the i-th GU can be expressed as follows $\mathbf{h}_k = \mathbf{G}_{S,k}^H + \mathbf{h}_{R,k}^H \Theta \mathbf{G}_{SR}$. Therefore, the received signal at the i-th GUs can be expressed as

$$y_i^{NOMA} = \mathbf{h}_i \sqrt{p_i} \mathbf{w}_i s_i + \sum_{j \neq i} \mathbf{h}_j \sqrt{p_j} \mathbf{w}_j s_j + n_i, \tag{4}$$

where p_i represents the power distributed to the i-th GUs, and n_i stands for AWGN. To mitigate interference among multiple GUs, we utilize the SIC-enabled technology. In this approach, the GUs are ordered according to their channel strengths, and the signal from the strongest channel is decoded first. The decoded signal is then subtracted from the received signal at the next stage, which improves the detection of the next strongest signal. For simplification, the channel coefficients of each ground-based station are listed as follows $|\mathbf{h}_K| \leq \ldots \leq |\mathbf{h}_2| \leq |\mathbf{h}_1|$ [55]. In this case, in order to ensure the smooth execution of SIC, the transmission energy of the satellite is constrained as follows

$$\Delta_{i,j} = p_i |\mathbf{h}_{i-1}|^2 - \sum_{j=1}^{i-1} p_j |\mathbf{h}_{i-1}|^2 \geq \rho_{\min}, \forall i \geq 2, \tag{5}$$

where the parameter $\rho_{\min} > 0$ denotes the gap in the decoded signals [56]. The instantaneous reachable rate of the NOMA algorithm on i-th GU can be obtained from the following formula:

$$R_i = B_i \log_2 \left(1 + \frac{|\mathbf{h}_i \mathbf{w}_i|^2 p_i}{\left|\mathbf{h}_j \mathbf{w}_j\right|^2 \sum_{j=1}^{j \neq i} p_j + \sigma_i^2} \right), \forall j \neq i, i \in K, \quad (6)$$

where B_i is the bandwidth distributed to the i-th GU.

4.2.2 Energy Dissipation Model for the UAV

This paper aims at a communication service based on static wing, which does not makes full use of the vertical movement characteristics and flight altitude difference of UAV. The purpose of the proposed model is to develop a new method for trajectory planning and hybrid beam synthesis. When the architecture is fully learned, factors such as 3D trajectory, vehicle user distribution, UAV distribution, and so on. In the design of the orbit, it can be assumed that all flight times are equal. When the time is T_K, the horizontal space coordinates of the n-th UAV can be expressed by $q_n^t \in Q$ with Q being the set of overflight trajectories. The center of the starting point and the last point are set in advance within a fixed service area. In this way, when the vertical space is not taken into account, $\left\{q_{S,n}^t, q_2^t, \ldots, q_n^t, \ldots, q_{N-1}^t, q_{D,n}^t\right\}$. In the vertical direction, the unmanned aerial vehicle operates at a fixed altitude in order to provide a more stable service.

On this basis, with reference to the mathematical model of energy consumption of rotor unmanned aerial vehicle during flight, the blade power consumption is set to $E_1(t) = \frac{\delta}{8} \rho_a f A^{\cdot\cdot 3} R^3 + n_E(t)$. Among them, f represents the resistance rate of the fuselage and rotor stability, rho_a represents the density of the gas, δ represents the drag coefficient of the cross section, A represents the area of the propeller, Ω represents the angle of the blade, and R represents the diameter of the whole blade. In addition, the project will use spin-wing drones to complement the two different business models of $n_E(t)$. In particular, this paper only takes the fixed flight in the static flight state as the research object, and does not make full use of its own vertical movement and flight altitude differences.

The total energy consumption of the UAV can be divided into two primary components: energy consumed for communication purposes and energy used for propulsion movements. It is worth noting that propulsive motion will consume at least 95% or more of the total energy consumption. This highlights the significant impact of propulsion-related energy expenditure on the UAV's overall energy usage. Propulsion movements involve the energy required to maintain the UAV's altitude and navigate along its trajectory. These activities demand substantial power to sustain the UAV's flight and enable it to move efficiently in the airspace. On the other hand, the energy consumed for communication purposes includes the power utilized for transmitting signals and maintaining wireless connectivity with the GUs. While this component is essential for enabling reliable communication, its contribution to the total energy consumption is relatively smaller compared to propulsion movements. Given this, our study focuses primarily on the energy consumption of UAVs during movement or hovering, which can be expressed as

$$E_k = E_m T_k v + E_s T_k + \varepsilon_e(t), \forall k, \tag{7}$$

where E_m and E_s are used to describe the moving energy of the machine and to transmit signals. In addition, the $\varepsilon_e(t)$ represents a collection of unconsidered errors generated over a period of time, influenced by the specific movement pattern of the UAV. These errors can arise from various sources such as modeling inaccuracies, environmental disturbances, and measurement noise. Notably, when the UAV is in hover, the engine can be turned off, resulting in no energy dissipation during this status.

4.2.3 Problem Formulation

Under the NOMA-enabled RIS-assisted SATINs system, the NOMA scheme enables the use of a single spatial direction to serve multiple GUs, even if their original channels are not aligned. This feature greatly simplifies the implementation of NOMA in the system. In NOMA, multiple users' signals are superimposed and transmitted in the same resource block. However, in a multi-antenna system like this, inter-user interference can become a significant issue due to the simultaneous transmission of multiple users' signals. To mitigate this interference and enhance system performance, beamforming technology is introduced. The beamforming technique allows for the shaping and steering of the transmitted signals towards specific directions, thereby reducing interference and improving signal quality.

The aim of this research contents is to maximize the total rate of the whole system while minimizing the energy consumption of UAVs. Here, $\mathbf{w} = [w_1, w_2, .., w_K]$ represents the active transmit beamforming vector, and $\mathbf{Q} = \left[\mathbf{q}_1, \mathbf{q}_2, ..., \mathbf{q}_n\right]^T$ represents the UAV's served trajectory. The mathematical expression for the optimization under this model was obtained over a longer running time, taking into account the effect of numerous constraints. On the premise of taking into account energy consumption, the optimization scheme of beamforming and track optimization is sought.

$$\max_{\boldsymbol{\Theta},\boldsymbol{w},Q} \frac{1}{T_k} \sum_{t=1}^{T_k} \sum_{k=1}^{K} \frac{R_k^t}{E_k} \tag{8a}$$

$$\text{s.t.} q_{S,n}^t = q_1^t, q_N^{T_k} = q_{D,n}^t, \forall k, \tag{8b}$$

$$\Delta_k^t \geq \rho_{\min}, p_k^t > 0, \forall k, \forall t, \tag{8c}$$

$$\sum_{k=1}^{K} p_k^t < P_{\max}, \forall k, \forall t, \tag{8d}$$

$$x_{\min} \leq x_k^t \leq x_{\max}, \tag{8e}$$

$$y_{\min} \leq y_k^t \leq y_{\max}, \tag{8f}$$

$$\left|\exp\left(j\theta_n^m\right)\right| = 1, \forall n, \forall m. \tag{8g}$$

where Eq. (8a) denoted as the optimisation objective of the proposed model through modelling. This requires that UAV can access more GUs during flight, so that the data transmission speed can be improved; therefore, the overall data transmission rate has been improved. Because UAV communication consumes less energy than sports behavior, hovering in areas where communication needs is a good choice. Emotional intelligence. Equation (8b) indicates the orbital design of the unmanned aircraft that decides to reach the start and end positions of the maximum service distance. The purpose of Eqs. (8c) and (8d) is to maintain the transmission energy and to ensure that the limit of power distribution is reached. The range of operations that UAV energy can support is limited to Eqs. (8e) and (8f). In addition, Eq. (8g) also represents the modulation limit that RIS must meet.

The optimization problem represented by Eq. (8) can be considered a non-convex problem due to several factors, as discussed in [57]. Firstly, the problem involves multiple parameters to be optimised in numerous transmission environments, namely ($\boldsymbol{\Theta}$, $\mathbf{w}$, $\mathbf{Q}$), which are more coupled and recursively related to each other's influence in the objective function. This interdependency makes the optimization problem complex and non-convex. Secondly, the system implementable rate R_k^t is not a continuous function due to the associated channel model coefficients at the angle of the signal emission due to the UAV position. This makes it difficult to find a global optimum as the objective function is discontinuous. Thirdly, when multiple UAVs are involved and move simultaneously, the complexity of the optimization problem increases significantly. In large-scale wireless communication networks, infinite variables are generated in continuous time, so even solving the subproblem of trajectory design becomes extremely challenging. To tackle these challenges, traditional algorithms often rely on suboptimal solutions obtained through alternating optimization techniques. These techniques involve iteratively optimizing subproblems while fixing other variables. For example, the advantage of the alternating iteration optimisation algorithm is that it can decompose the multi-objective problem into a series of independent and repeatable subproblems, and at the same time, it can also avoid dealing with high-dimensional problems directly and reduce the difficulty of solving. However, it should be noted that there may be conflicts or interdependence between the subproblems, which requires appropriate adjustment and coordination in selecting the subproblems and updating the current solution. To solve this problem, this project intends to draw lessons from the advanced multi-vector DDPG (MV-DDPG) model. In this project, an optimal mathematical model based on the mathematical model is proposed, and an efficient and energy-saving solution method is proposed through the mathematical modeling of the optimal problem. This framework formulates the optimization problems and aims to find feasible solutions for $\mathbf{w}$ and $\boldsymbol{\Theta}$ to enhance energy efficiency.

4.3 Proposed Optimized Approach

Many problems such as UAV trajectory design, transmit beamforming, and RIS configuration are transformed into a multi-objective optimization problem. This project intends to study the multi-vector DDPG (MV-DDPG) method with high sampling efficiency. The considered MV-DDPG algorithm framework is well-suited for solving non-linear problems with multiple objectives and has been successfully used in a variety of applications.

4.3.1 MDP Formulation

The optimal transmission beamforming scheme is represented as an optimization problem in Eq. (8). This problem is a successive decision process over time, meaning that decisions are made based on current environmental interactions. To solve this optimization problem, the project utilizes the Markov decision process (MDP) method. The MDP method is applied to transform the optimization problem into a tuple consisting of five components: $\langle S, A, R, T, \gamma \rangle$, standing for the state space, action space, reward function, transformation strategy, and discount factor. Through the modeling of the optimal problem, we can use a variety of methods such as reinforcement learning to find the optimal transmission scheme.

To achieve the best practices strategy, denoted as $\pi^* (s, a)$, the learning agent must maximizes the state-value function $Q_\pi (s)$, which represents the expected cumulative discounted reward that an agent can achieve by following a specific policy π from a given state s. This optimization process involves evaluating the value of each state-action pair and iteratively updating the policy based on the obtained rewards and future expectations. It is worth noting that the quality of the learned policy depends heavily on the quality and quantity of the training data. Therefore, generating sufficient and diverse training data is crucial for achieving high-quality performance. Moreover, incorporating additional information such as expert knowledge or prior experience can further enhance the learning process and improve the quality of the learned policy. This involves evaluating the value of each state-action pair and iteratively updating the policy based on the obtained rewards and future expectations. The state-value function provides a measure of the long-term desirability of a particular state under a given policy. By estimating and maximizing this function, the learning agent can determine which states are more valuable and make better decisions to achieve higher cumulative rewards over time. This Q-value function satisfies the Bellman equation, which expresses the relationship between the value of a state-action pair and the value of the subsequent state-action pairs.

By iteratively solving the Bellman equation, the learning agent can estimate and update the Q-values to ultimately find the optimal policy for maximizing long-term rewards. In addition to the Bellman equation, another important concept in reinforcement learning is the Bellman expectation equation. This equation expresses the expected value of the Q-function for all possible actions in a given state and is used in algorithms such as policy iteration and value iteration. Mathematically, the Bellman expectation equation can be represented as

$$\begin{aligned} V_\pi(s) &= \mathbb{E}_\pi\left[G_t \mid S_t = s\right] \\ &= \mathbb{E}_\pi\left[\sum_{k=0}^{\infty} \gamma^k R_{t+k+1} \mid S_t = s\right] \\ &= \sum_{a \in A} \pi(a \mid s)\left(R_s^a + \gamma \sum_{s' \in S} P_{ss'}^a Q_\pi(s')\right), \end{aligned} \tag{9}$$

where $R_s^a = \mathbb{E}\left[R_{t+1} \mid S_t = s, A_t = a\right]$ denotes the reward obtained when the agent is in state s, and $P_{ss'}^a$ represents the probability of transitioning from state s to the next state s'. The state transition probability matrix provides information on the likelihood of moving to different states based on the current state and action taken. It's worth noting that in this MDP model, there is a central controller acting as an agent to explore the unknown wireless environment. The central controller interacts with the environment, makes decisions based on the current state, and receives rewards accordingly. The construction of MDP in the considered DRL framework can be in detail expressed as:

(1) *State Space*: This comprehensive state space provides a holistic view of the system's current state, facilitating effective decision-making and control. Among them, the parameters of the state space $s^{(t)}$ include four parts: the transmission power strategy, the regional point where UAV is located q_n^t, the last time selected action, and the relevant channel matrix $\mathbf{h}_k$, as well as the associated channel matrix $\mathbf{G}_r$. The state space is as follows

$$s^{(t)} = \left\{p_k, q_n[t], a^{[t-1]}, \mathbf{h}_k[t], \mathbf{G}_r[t]\right\}. \tag{10}$$

When the positioning of the unmanned aerial vehicle is completed, the hybrid beamforming scheme is carried out. On this basis, a novel beamforming method is proposed. This method not only improves the communication quality, but also provides a reference for the positioning of UAV. This project intends to use the cross-optimization method to overcome the asymmetry of space-based and inter-satellite information and make more users participate in its accessible business. By continuously optimizing the beamforming techniques and adjusting the UAV position, the algorithm aims to maximize the available service time for users in different areas.

(2) *Action Space*: The proposed framework for the MDP problem is unique due to the inclusion of both real and imaginary parts in the RIS passive reflection component. This leads to a mixture of discrete and continuous action spaces, making the problem more complex [58]. To address this challenge, certain modifications are made to the action space. The UAV maneuver movement direction is divided into four options: leftward, forward, rightward, and backward. This discretization allows for easier decision-making

regarding the movement of the UAV. Similarly, the satellite transmission power is also discretized to simplify the optimization process. Next, the RIS phase shift at the m-th element, denoted as θ_n^m is considered. The phase shift is chosen from a set of discrete values given by $\theta_n^m \in \{1, \Delta\theta, ..., (L-1)\Delta\theta\}$ This discrete selection of phase shifts allowing for more manageable computations and practical implementation. Finally, the transmit beamforming matrix for the satellite, denoted as $\mathbf{w}[t] = \left[w_1^t, w_2^t, ..., w_K^t\right]^T$ is included in the action space. The action space can be expressed as

$$a^{(t)} = \left\{q_n^t, \theta_n^m, \mathbf{w}[t]\right\}, \forall t, \forall m, \forall n. \tag{11}$$

(3) *Reward Function*: To ensure that the guided learning process aligns with this objective, the reward function needs to be designed accordingly. Thus, we set a penalty for the constraints defined in Eqs. (8b)–(8e). Specifically, if any of these constraints are not satisfied, the penalty terminates the set of constraints. This approach ensures that the system operates within the specified constraints and maximizes the achievable rate. The reward function can be defined as

$$R^{(t)} = \begin{cases} -W_c, & \text{if } S_m = NS, \\ R[t], & \text{otherwise} \end{cases}. \tag{12}$$

where NS denotes the concept of a negative state, can be introduced to represent a state that does not satisfy any of the constraints. W_c denotes the concept of a sufficiently large set of normal quantities. By including W_c in the optimization process, we ensure that the guided learning process takes into account the constraints and avoids actions that would lead to constraint violations. This approach promotes the maximization of the total achievable rate while maintaining system stability and adherence to the specified constraints.

The proposed MDP model involves uncertainty regarding the UAV location and RIS phase shift configuration, making it difficult to determine the state space transition probability matrix $P_{ss'}^a$. On the other side, the DRL method can effectively regulate the behavior of the mobile terminal on the premise of accurate positioning. In DRL-enabled authorization mode, agents make behavior decisions according to their own state and receive feedback information. By constantly interacting with their surroundings, agents learn how to maximize cumulative revenue over a period of time.

4.3.2 MV-DDPG Algorithm Description

Figure 4 illustrates the architecture of the MV-DDPG involved neural network, which comprises two distinct components: the actor network and the critic network. These networks play crucial roles in the reinforcement learning process by enabling the generation of actions and estimating their corresponding values. The actor network takes as input the current state and outputs a set of actions. It acts as the policy function, mapping states to appropriate actions based on the learned behavior. This allows the agent to make informed decisions on how to interact with the environment and navigate towards desirable outcomes. On the other hand, the critic network assesses the quality or value of the selected actions. It takes as input both the current state and the actions generated by the actor network, and provides an estimate of their value. Both networks utilize a dual network structure and the soft-update strategy, and they are updated using target network, $\pi\left(\theta_a^{(target)}|s^{(t)}\right)$ and $Q\left(\theta_c^{(target)}|s^{(t)}, a^{(t)}\right)$, respectively. During the training process, the actor network learns to generate optimal actions based on the current state, while the critic network learns to estimate the Q-value of the selected actions. In the learning algorithm, a learning algorithm based on random gradient is used to modify the relationship between the role and the evaluator. The results of the participant network define the basic principles for establishing the correspondence between the situation and the behavior, while the results of the critic network can evaluate the value of the behavior. The weights and preference coefficients between evaluators and evaluators are expressed as θ_a and θ_c.

The training of the critic network in MV-DDPG involves updating its parameters to minimize the mean square error (MSE) loss between the estimated Q-value and the target Q-value. The target Q-value is calculated using the Bellman equation, which recursively estimates the expected future reward. By minimizing the MSE loss between the estimated Q-value and the target Q-value, the critic network can learn to accurately estimate the Q-value function, even in the presence of multiple objectives.

$$\theta_c^{(t+1)} = \theta_c^{(t)} - \mu_c \nabla_{\theta_c^{(train\,)}} \ell\left(\theta_c^{(train\,)}\right), \tag{13}$$

$$\begin{aligned} \ell\left(\theta_c^{(train)}\right) = & \left(r^{(t)} + \gamma Q\left(\theta_c^{(target)} \mid s^{(t+1)}, a'\right)\right. \\ & \left.- Q\left(\theta_c^{(train)} \mid s^{(t)}, a^{(t)}\right)\right)^2, \end{aligned} \tag{14}$$

where μ_c is the learning rate. It determines the step size during the parameter update. The action output a' represents the actions generated by the actor network for a given state. The update gradient $\nabla_{\theta_c^{(train)}} \ell\left(\theta_c^{(train)}\right)$ signifies the gradient of the loss function. In the realm of actor-critic networks, it is customary to begin the learning process by initializing the weights. These weights play a crucial role in determining the overall performance and effectiveness of both the actor and critic networks. The actor network is responsible for making decisions and taking actions based on the given environment and input. Similarly, the critic network evaluates and provides feedback on the actor's performance. Therefore, initializing the weights of both networks is of utmost importance in ensuring their optimal functionality. To initialize these weights, a truncated normal distribution centered at 0 is commonly employed. This distribution considers the statistical characteristics of the weights and ensures that they are spread out in a manner that allows for efficient learning and exploration within the network. It allows for a smoother convergence during the learning process and helps avoid issues such as vanishing or exploding gradients.

The training of the actor network in MV-DDPG involves updating its parameters using the policy gradient as follows

$$\theta_a^{(t+1)} = \theta_a^{(t)} - \mu_a \nabla_a Q\left(\theta_c^{(target)} \mid s^{(t)}, a'\right) \nabla_{\theta_a^{(train)}} \pi\left(\theta_a^{(train)} \mid s^{(t)}\right), \tag{15}$$

where μ_a denotes the learning rate. The update of the actor network's parameters involves two gradient terms: $\nabla_a Q\left(\theta_c^{(target)} \mid s^{(t)}, a'\right)$ and $\nabla_{\theta_a^{(train)}} \pi\left(\theta_a^{(train)} \mid s^{(t)}\right)$. The first term in the above part represents the gradient of the target critic network with respect to its parameter. This gradient provides guidance for updating the actor network parameters in the right direction for implementation actions, ultimately optimizing the Q-value function. By incorporating this gradient, the actor network can adjust its parameters to improve its performance and make more informed decisions based on the desired objective. The second term represents the gradient of the actor network's policy with respect to its parameter. This gradient guides the update of the actor network parameters based on the current state, influencing the selection of the next action. To make the training process smoother, we use a targeted network that contains an object's network $\theta_a^{(target)}$ and target critic network $\theta_c^{(target)}$. During initialization, the weight of each learning network is obtained by copying the relevant parameters of each training network.

This framework introduces a strategy to reduce the dimensionality of the input space and enhance training efficiency by selecting a subset of system information closely related to the decision-making process. Instead of using all available system information as input to the neural network, the framework extracts a small subset of relevant information to form the state vector. This selective approach allows the network to focus on the most critical factors affecting the decision at hand, reducing the complexity of the input space and improving the learning process. To capture this preference, a linear weighting approach is adopted. In the context of optimizing the performance of a RIS-UAV-assisted system, a weight vector $\mathbf{v} = \left[v_{R_K}, v_{E_k}\right]$ is used to represent the weights assigned to different objectives or constraints. Specifically, the weight v_{R_K}, is associated with the rate optimization objective, while the weight v_{E_K} is associated with the energy constraint of the UAV. The weight vector $\mathbf{v}$ allows for balancing the trade-off between different objectives or constraints in the optimization process. By assigning appropriate weights to each objective or constraint, the algorithm can prioritize certain objectives over others and find a suitable compromise solution that best satisfies the given optimization criteria. To compute the scalar-type reward suitable for the network input, the reward vector is multiplied element-wise by the weight vector. This multiplication captures the trade-off between different objectives and generates a single scalar reward that reflects the overall desirability of the chosen action. It enables the learning agent to make informed decisions that balance different optimization targets, leading to improved system efficiency and effectiveness. Thus, the target value y_t is computed as

$$y_t = \mathbf{r}_t \mathbf{w}^T + \gamma \max_{a'} Q\left(\theta_a^{(target)} | s^{(t+1)}, a'\right). \tag{16}$$

To optimize the main critic network in the actor-critic algorithm, temporal difference learning is employed. This technique involves computing the discrepancy between the value of the Q-function outputted by the network and the estimated value of the objective function. This calculation serves as a crucial step in assessing the performance of the network and optimizing its predictions. By comparing these values, the network can assess its performance and make adjustments to optimize its predictions. Minimizing this loss function is crucial for training the main critic network effectively. To achieve this, a gradient descent approach is utilized. By taking the gradient of the loss function with respect to the weights of the network, the algorithm determines the direction in which the weights should be adjusted to minimize the error. The weights are then updated iteratively using the gradient descent update rule, which involves subtracting a fraction of the gradient from the current weights. This iterative process continues until convergence is achieved or a predefined stopping criterion is met. Through this training procedure, the main critic network gradually learns to approximate the true value of the Q-function more accurately, leading to improved performance and better policy evaluation.

By iteratively adjusting the network parameters in the direction that reduces the loss, the network gradually improves its ability to estimate the value of actions in relation to the given objectives. To ensure that the training process explores the full space of possible actions and avoids getting stuck in local optimal value, exploration strategies are incorporated into the actor policy. These strategies introduce a certain level of randomness in the selection of actions during decision-making. During the execution of an action in the actor-critic algorithm, a random amount of operations is chosen at each time step. The distribution of this random variable is characterized by an expectation and variance denoted by $\varepsilon_i \sigma^2$. Here, the parameter ε_i can be adjusted to control the impact of the randomness in action selection during training. This helps to prevent the network from becoming too focused on a limited set of actions and promotes the discovery of potentially better solutions.

4.4 Experimental Settings and Results

Through studies in multiple condition designs, the application effect of this method under various performance indexes is analyzed. Additionally, the randomly generated channel matrix $\mathbf{h}_k$ and $\mathbf{G}_r$ are used to verify the effectiveness of the proposed method. The algorithm proposed in this paper is compared with the existing algorithms abroad, which shows that the algorithm has excellent performance in each index. The specific simulation parameters are presented in Table 1, providing detailed information on the configuration and settings used for the performance demonstration [59].

In the initial phase of optimization for each iteration of the computational process, the UAV initializes a 2D trajectory randomly within a predefined area. It should be emphasized

Table 1 Simulation parameters setting

Simulation system parameters	Value
Satellite	GEO
Satellite antenna gain	48 dB
Frequency	2.4 GHz
Bandwidth	$B = 15\,\text{MHz}$
Noise power spectral density	–169 dBm/Hz
Variance of AWGN	1
Noise temperature	$300^{\circ} K$
UAV flying height	$\mathbf{H}_0 = 1000\,\text{m}$
UAV flying speed	$V_H = 10\,\text{m/s}$
Simulation DNN parameters	Value
Amount of time slots	100
Learning rate	0.002
Size of experience replay buffer	50000
Amount of episode	5000
Optimizer Type	AdamOptimizer
Activation function	ReLu
Soft target update parameter	0.001
Amount of step in per-episode	40000
Amount of experience in the mini-batch	64

that the UAVs are meticulously engineered to ensure coverage of an area with a radius of at least 10 km, a critical requirement for effectively serving the assigned service area. This comprehensive approach not only guarantees the establishment and optimization of the communication network but also significantly contributes to enhancing the overall effectiveness of the system. The meticulous design and deployment of UAVs exemplify a strategic commitment to achieving robust and efficient communication infrastructure in the pre-designated flight service area.

The proposed MV-DDPG algorithm in this paper employs a fully connected neural network structure for both the actor and critic networks. Additionally, modular normalization components are incorporated into the network architecture, as illustrated in Fig. 4. To train the policy and Q-networks, AdamOptimizers are utilized, and the parameters of each deep neural network are randomly initialized following a zero-mean normal distribution. This initialization strategy helps ensure that the networks start with diverse initial weights, enabling them to explore a wide range of possible solutions during the training process. In order to evaluate the effectiveness of the proposed framework, the paper also investigates the utilization of the NOMA protocol and examines the system performance under the RIS scenario.

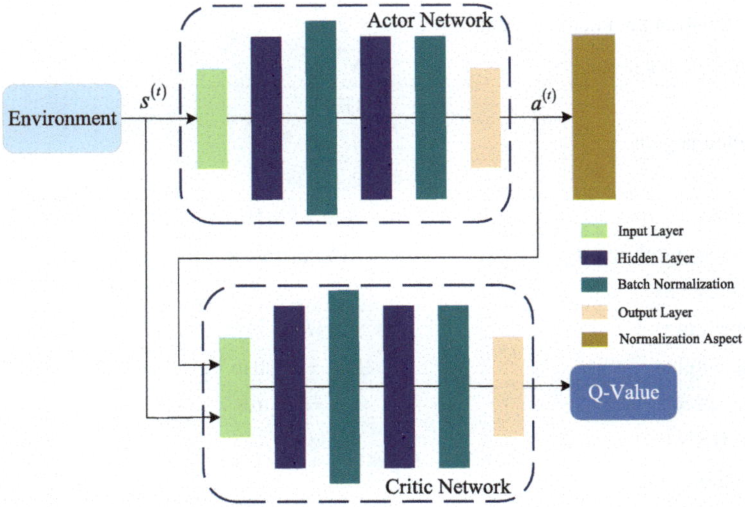

Fig. 4 DNN structures considered for actor networks and critic networks

By considering these additional variables, the study provides a comprehensive analysis of the proposed algorithm's capabilities and its potential application in real-world scenarios. The combination of the fully connected neural network structure, modular normalization components, and the examination of different protocol options and scenarios contributes to the novelty and effectiveness of the proposed MV-DDPG algorithm. Specifically, the evaluation of the network performance includes three key benchmarks:

- **Random assignment strategy**: At each time slot, this random generation of phase shift values is an essential component of the proposed scheme, enabling the RIS to effectively enhance the communication performance in a dynamic and flexible manner.
- **Without RIS**: This means that the RIS technology is not utilized in the system design and operation.

Our system modeling assumes that the UAV is at a fixed altitude, so we use a two-dimensional orbit to show various optimization purposes, as shown in Figs. 5 and 6. It is found that the RIS technology with optimal phase shift ratio can expand the communication range in a large range and save the minimum thrust energy consumption under certain power constraints. In addition, if there is without RIS system, the UAV must find a suitable place to connect the communication link with the intended GU, resulting in low energy-efficient performance. By comparing the results obtained using the three benchmarks with those obtained using the proposed MV-DDPG algorithm, we demonstrate the superiority of our approach in optimizing UAV communication systems. Our approach not only outperforms traditional OMA protocols but also shows significant improvements over NOMA

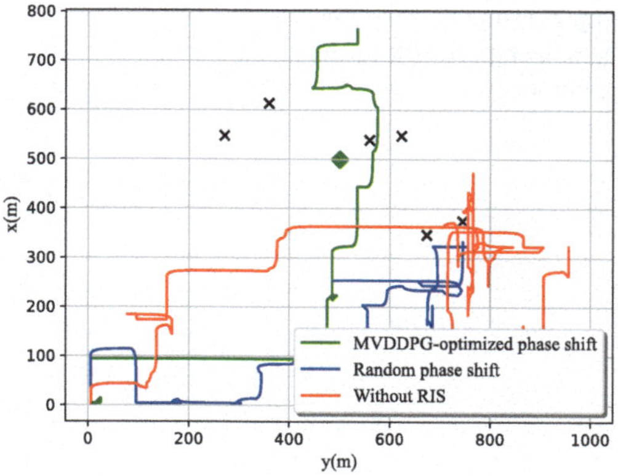

Fig. 5 2D Trajectory-Coverage area

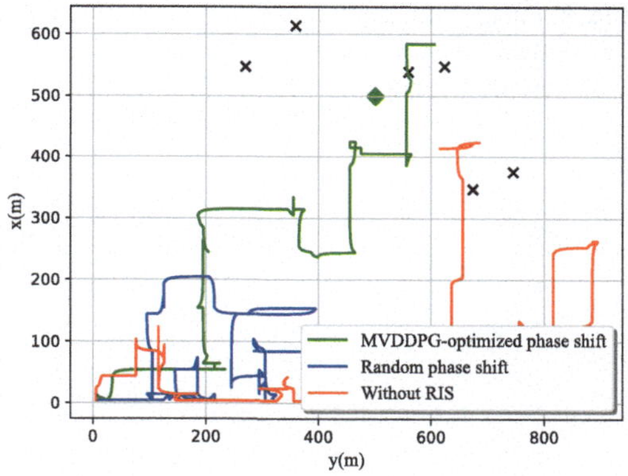

Fig. 6 2D Trajectory-Energy constraint

and RIS scenarios. The effectiveness of the MV-DDPG algorithm is evident in achieving optimal performance even when dealing with dynamic and uncertain environments. Overall, the proposed algorithm provides a promising solution for optimizing UAV communication systems.

In Figs 7, 8, 9 and 10, we present the cumulative distribution functions (CDF) of a wide range of performance indicators for three different communication systems. The results demonstrate that RIS-assisted communication with optimized phase shifts achieves superior performance in various wireless scenarios. This is attributed to the random nature of the phase shifts and the UAV's dynamic location, which poses challenges in aligning signals and leads to performance degradation and reduced communication energy efficiency. Furthermore, we evaluate the performance of the NOMA protocol in combination with RIS technology under different scenarios. We consider the following three benchmarks

- **NOMA-enabled Without RIS**: In this scheme, the UAV does not carry RIS. The communication relies solely on the direct link from the satellite to the GUs, utilizing the

Fig. 7 Data rate performance demonstration with various solutions

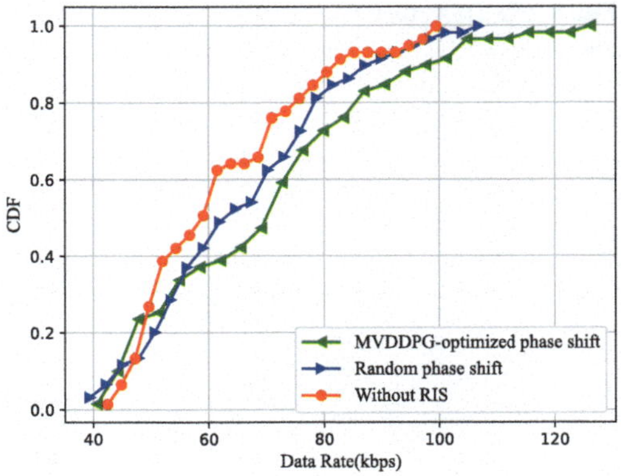

Fig. 8 Energy-efficiency performance demonstration with various solutions

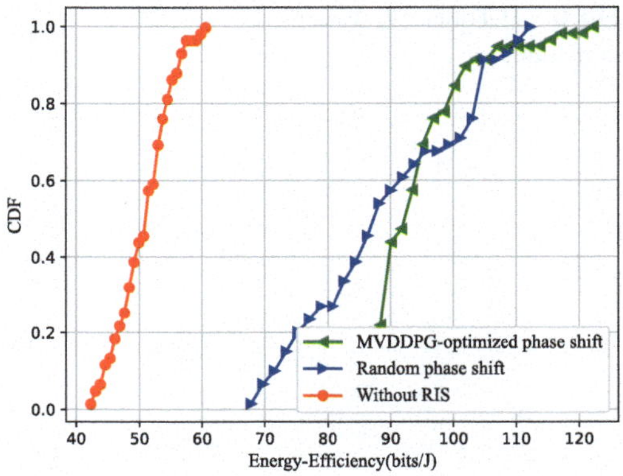

Fig. 9 Throughput performance demonstration with various solutions

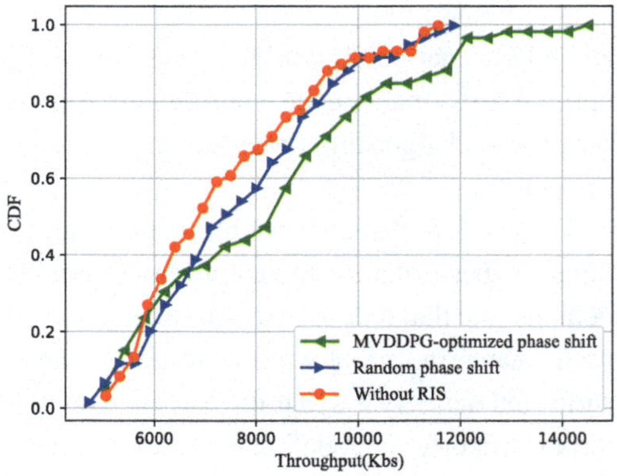

NOMA protocol as the user access scheme. This scenario allows for an assessment of the performance of the NOMA protocol without the influence of RIS technology.

- **Without RIS-OMA**: The scheme is similar to the previous one, but it employs the OMA protocol as the user access scheme instead of NOMA. This comparison enables a direct

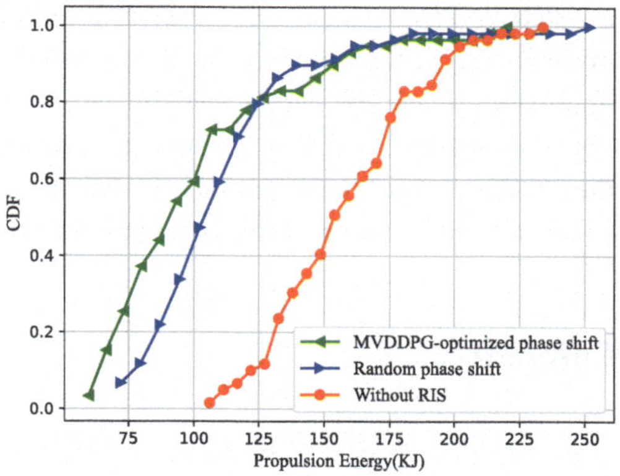

Fig. 10 Propulsion energy performance demonstration with various solutions

evaluation of the performance difference between NOMA and OMA protocols in the absence of RIS technology.

- **Optimal phase shift-OMA**: In this scenario, the active and passive beamforming designs are optimized using the MV-DDPG framework. However, unlike the previous scenarios, the user access scheme in this case follows the OMA protocol. This scenario allows for an investigation of the impact of optimized beamforming on system performance under the OMA protocol.

The proposed MV-DDPG algorithm framework in this paper can arbitrarily extend the optimisation dimensions to suit the transmission conditions under different scenarios. Moreover, the above analysis clearly illustrates that the sum rate experiences a progressive increase as the number of reflective elements in the RIS grows. This trend can be attributed to several factors. Firstly, a larger metasurface enables the formation of more reflective elements, thereby enhancing the passive beamforming gain and improving the overall system performance. Additionally, the application of the NOMA protocol further enhances the spectral efficiency, contributing to the observed increase in the sum rate. This insightful analysis provides valuable insights into the potential benefits and performance improvements achievable by employing the MV-DDPG algorithm in various scenarios of UAV communication systems. The deployment of this algorithm holds great promise for optimizing communication efficiency and achieving superior performance in UAV-based networks.

4.5 Conclusions

The primary focus of this paper was to explore the characteristics of the RIS-assisted SATIN system and address the transmission challenges they encounter. This exploration ultimately led to the proposition of enhancing system performance through the introduction of RIS. The paper also delved into the presentation of research directions and challenges for RIS-assisted SATINs system. Moreover, it introduced a NOMA-assisted system performance optimization scheme within the RIS-assisted SATINs system, aiming to identify a joint optimization solution for MV-DPPG based on DRL while taking into account multiple objective-related weighting parameters. The simulation results convincingly demonstrated the effectiveness of the proposed scheme, thereby providing a valuable theoretical reference solution for the industrial implementation of RIS.

References

1. "Cisco global networking trends: Executive summary"; https://www.cisco.com/us/solutions/executive-perspectives/global-networking-trends-2022 html, accessed Nov. 2022.
2. K. Guo, X. Li, M. Alazab, R. H. Jhaveri and K. An, "Integrated satellite multiple two-way relay networks: Secrecy performance under multiple eves and vehicles with non-ideal hardware," *IEEE Trans. Intelligent Veh.*, vol. 8, no. 2, pp. 1307–1318, 2023.
3. Y. Sun et al., "Energy-efficient hybrid beamforming for multilayer RIS-assisted secure integrated terrestrial-aerial networks," *IEEE Trans. Commun.*, vol. 70, no. 6, pp. 4189–4210, 2022.
4. M. Renzo, Zappone, M. Debbah, M. S. Alouini, S. Tretyakov, "Smart radio environments empowered by reconfigurable intelligent surfaces: How it works, state of research, and the road ahead," *IEEE J. Sel. Areas Commun.*, vol. 38, no. 11, pp. 2450–2525, Nov. 2020.
5. X. Pang, M. Sheng, N. Zhao, J. Tang, D. Niyato and K. -K. Wong, "When UAV meets IRS: Expanding air-ground networks via passive reflection," *IEEE Wireless Commun.*, vol. 28, no. 5, pp. 164–170, 2021.
6. C. Wang *et al.*, "Covert communication assisted by UAV-IRS," *IEEE Trans. Commun.*, vol. 71, no. 1, pp. 357–369, 2023.
7. L. Yang, Y. Yang, D. B. d. Costa and I. Trigui, "Outage probability and capacity scaling law of multiple RIS-aided networks," *IEEE Wireless Commun. Lett.*, vol. 10, no. 2, pp. 256–260, Feb. 2021.
8. F. Zhou, X. Li, M. Alazab, R. H. Jhaveri and K. Guo, "Secrecy performance for RIS-based integrated satellite vehicle networks with a UAV relay and MRC eavesdropping," *IEEE Trans. Intelligent Vehi.*, vol. 8, no. 2, pp. 1676–1685, 2023.
9. S. Ouahouah, M. Bagaa, J. Prados-Garzon and T. Taleb, "Deep-reinforcement-learning-based collision avoidance in UAV environment," *IEEE Internet Things J.*, vol. 9, no. 6, pp. 4015–4030, 2022.
10. C. Niephaus *et al.*, "QoS provisioning in converged satellite and terrestrial networks: A Survey of the state-of-the-art," *IEEE Commun. Surveys Tutorials*, vol. 18, pp. 2415–41, Fourth Quarter, 2016.
11. N. Agrawal, A. Bansal, K. Singh, C. -P. Li and S. Mumtaz, "Finite block length analysis of RIS-assisted UAV-based multiuser IoT communication system with non-linear EH," *IEEE Trans. Commun.*, vol. 70, no. 5, pp. 3542–3557, May. 2022.
12. S. Liu, Z. Gao, J. Zhang, M. D. Renzo, and M.-S. Alouini, "Deep denoising neural network assisted compressive channel estimation for mmWave intelligent reflecting surfaces," *IEEE Trans. Veh. Technol.*, vol. 69, no. 8, pp. 9223–9228, 2020.
13. J. An, C. Xu, L. Gan and L. Hanzo, "Low-complexity channel estimation and passive beamforming for RIS-assisted MIMO systems relying on discrete phase shifts," *IEEE Trans. Commun.*, vol. 70, no. 2, pp. 1245–1260, Feb. 2022.

14. P. S. Bithas, V. Nikolaidis, A. G. Kanatas and G. K. Karagiannidis, "UAV-to-ground communications: Channel modeling and UAV selection," *IEEE Trans. Commun.*, vol. 68, no. 8, pp. 5135–5144, 2020.
15. S. Xu *et al.*, "Envisioning intelligent reflecting surface empowered space-air-ground integrated network," *IEEE Netw.*, vol. 35, no. 6, pp. 225–232, 2021.
16. C. Huang, G. Chen, J. Tang, P. Xiao and Z. Han, "Machine-learning-empowered passive beamforming and routing design for multi-RIS-assisted multihop networks," *IEEE Internet Things J.*, vol. 9, no. 24, pp. 25673–25684, 15 Dec.15, 2022.
17. Z. Yang, Y. Liu, Y. Chen and N. Al-Dhahir, "Machine learning for user partitioning and phase shifters design in RIS-aided NOMA networks," *IEEE Trans. Commun.*, vol. 69, no. 11, pp. 7414–7428, 2021.
18. K. M. Faisal and W. Choi, "Machine learning approaches for reconfigurable intelligent surfaces: A survey," *IEEE Access*, vol. 10, pp. 27343–27367, 2022.
19. X. Liu, Y. Liu and Y. Chen, "Machine learning empowered trajectory and passive beamforming design in UAV-RIS wireless networks," *IEEE J. Sel. Areas in Commun.*, vol. 39, no. 7, pp. 2042–2055, 2021.
20. H. Zhou, C. She, Y. Deng, M. Dohler and A. Nallanathan, "Machine learning for massive industrial internet of things, " *IEEE Wireless Commun.*, vol. 28, no. 4, pp. 81–87, 2021, https://doi.org/10.1109/MWC.301.2000478.
21. K. Arulkumaran, M. P. Deisenroth, M. Brundage and A. A. Bharath, "Deep reinforcement learning: A brief survey," *IEEE Signal Processing Mag.*, vol. 34, no. 6, pp. 26–38, 2017.
22. I. Ahmad, R. Narmeen, Z. Becvar and I. Guvenc, "Machine learning-based beamforming for unmanned aerial vehicles equipped with reconfigurable intelligent surfaces," *IEEE Wireless Commun.*, vol. 29, no. 4, pp. 32–38, 2022.
23. Y. Li. (2017). Deep reinforcement learning: An overview. arXiv. [Online]. Available: https://arxiv.org/abs/1701.07274
24. V. Mnih *et al.*, "Human-level control through deep reinforcement learning," *Nature*, vol. 518, no. 7540, pp. 529–533, 2015.
25. N. C. Luong *et al.*, "Applications of deep reinforcement learning in communications and networking: A survey," *IEEE Commun. Surveys Tuts.*, vol. 21, no. 4, pp. 3133–3174, 4th Quart., 2019.
26. T. Lin and Y . Zhu, "Beamforming design for large-scale antenna arrays using deep learning," *IEEE Wireless Commun. Lett.*, vol. 9, no. 1, pp. 103–107, 2020.
27. F. Zhou, G. Lu, M. Wen, Y.-C. Liang, Z. Chu, and Y. Wang, "Dynamic spectrum management via machine learning: State of the art, taxonomy, challenges, and open research issues," *IEEE Netw.*, vol. 33, no. 4, pp. 54–62, 2019.
28. H. Huang, Y. Song, J. Y ang, G. Gui, and F. Adachi, "Deep learning-based millimeter-wave massive MIMO for hybrid precoding," *IEEE Trans. Veh. Technol.*, vol. 68, no. 3, pp. 3027–3032, Mar. 2019.
29. X. Li and A. Alkhateeb, "Deep learning for direct hybrid precoding in millimeter wave massive MIMO systems," 2019, *arXiv:* 1905.13212. [Online]. Available: http://arxiv.org/abs/1905.13212
30. Y. Zhou, F. Zhou, Y. Wu, R. Q. Hu, and Y. Wang, "Subcarrier assignment schemes based on Q-learning in wideband cognitive radio networks," *IEEE Trans. Veh. Technol.*, vol. 69, no. 1, pp. 1168–1172, 2020.
31. F. B. Mismar, B. L. Evans, and A. Alkhateeb, "Deep reinforcement learning for 5G networks: Joint beamforming, power control, and interference coordination," *IEEE Trans. Commun.*, vol. 66, no. 3, pp. 1581–1592, 2020.
32. R. Shafin *et al.*, "Self-tuning sectorization: Deep reinforcement learning meets broadcast beam optimization," 2019, *arXiv:* 1906.06021. [Online]. Available: http://arxiv.org/abs/1906.06021

33. X. Liu, Y. Liu, Y. Chen and L. Hanzo, "Trajectory design and power control for multi-UAV assisted wireless networks: A machine learning approach," *IEEE Trans. Veh. Technol.*, vol. 68, no. 8, pp. 7957–7969, Aug. 2019.
34. K. Feng, Q. Wang, X. Li and C. -K. Wen, "Deep reinforcement learning based intelligent reflecting surface optimization for MISO communication systems," *IEEE Wireless Commun. Lett.*, vol. 9, no. 5, pp. 745–749, 2020.
35. K. Guo *et al.*, "Performance analysis of hybrid satellite-terrestrial cooperative networks with relay selection," *IEEE Trans. Vehi. Technol.*, vol. 69, no. 8, pp. 9053–9067, 2020.
36. K. Guo *et al.*, "Deep reinforcement learning and NOMA-based multi-objective RIS-assisted IS-UAV-TNs: Trajectory optimization and beamforming design" *IEEE Trans. Intell. Transp. Syst.*, vol. 24, no. 9, pp. 10197–10210, 2023.
37. C. Huang, R. Mo and C. Yuen, "Reconfigurable intelligent surface assisted multiuser MISO systems exploiting deep reinforcement learning," *IEEE J. Sel. Areas Commun.*, vol. 38, no. 8, pp. 1839–1850, 2020.
38. T. Cover, "Broadcast channels," *IEEE Trans. Inf. Theory*, vol. 18, no. 1, pp. 2–14, 1972.
39. S. Vanka, S. Srinivasa, Z. Gong, P. Vizi, K. Stamatiou and M. Haenggi, "Superposition coding strategies: Design and experimental evaluation," *IEEE Trans. Wireless Commun.*, vol. 11, no. 7, pp. 2628–2639, 2012.
40. X. Yue, Z. Qin, Y. Liu, S. Kang and Y. Chen, "A unified framework for non-orthogonal multiple access," *IEEE Trans. Commun.*, vol. 66, no. 11, pp. 5346–5359, 2018.
41. Z. Ding, X. Lei, G. K. Karagiannidis, R. Schober, J. Yuan, and V. Bhargava, "A survey on non-orthogonal multiple access for 5G networks: Research challenges and future trends," *IEEE J. Sel. Areas Commun.*, vol. 35, no. 10, pp. 2181–2195, 2017.
42. Z. Ding *et al.*, "Application of non-orthogonal multiple access in LTE and 5G networks," *IEEE Commun. Mag.*, vol. 55, no. 2, pp. 185–191, 2017.
43. X. Li *et al.*, "Residual transceiver hardware impairments on cooperative NOMA networks," *IEEE Trans. Wireless Commun.*, vol. 19, no. 1, pp. 680–695, 2020.
44. X. Li *et al.*, "Secrecy analysis of ambient backscatter NOMA systems under I/Q imbalance," *IEEE Trans. Veh. Technol.*, vol. 69, no. 10, pp. 12286–12290, 2020.
45. Y. -L. Lan, F. Liu, W. W. Y. Ng, M. Gui and C. Lai, "Multi-objective two-echelon city dispatching problem with mobile satellites and crowd-shipping," *IEEE Trans. Intell. Transp. Syst.*, vol. 23, no. 9, pp. 15340–15353, 2022.
46. J. Watts, F. Van Wyk, S. Rezaei, Y. Wang, N. Masoud and A. Khojandi, "A dynamic deep reinforcement learning-bayesian framework for anomaly detection," *IEEE Trans. Intell. Transp. Syst.*, vol. 23, no. 12, pp. 22884–22894, 2022.
47. M. Luo *et al.*, "Fleet rebalancing for expanding shared e-mobility systems: A multi-agent deep reinforcement learning approach," *IEEE Trans. Intell. Transp. Syst.*, vol. 24, no. 4, pp. 3868–3881, 2023.
48. C. Zhong *et al.*, "Deep reinforcement learning-based optimization for IRS-assisted cognitive radio systems," *IEEE Trans. Commun.*, vol. 70, no. 6, pp. 3849–3864, 2022.
49. X. Liu, Y. Liu, and Y. Chen, "Machine learning empowered trajectory and passive beamforming design in UAV-RIS wireless networks," *IEEE J. Sel. Areas Commun.*, vol. 39, no. 7, pp. 2042–2055, 2021.
50. M. K. Arti, "Channel estimation and detection in satellite communication systems," *IEEE Trans. Veh. Tech.*, vol. 65, no. 12, pp. 10173–10179, 2016.
51. K. Guo, M. Lin, B. Zhang, W. Zhu, J. Wang, and T. A. Tsiftsis, "On the performance of LMS communication with hardware impairments and interference," *IEEE Trans. Commun.*, vol. 67, no. 2, pp. 1490–1505, 2019.
52. G. Zheng, P. D. Arapoglou, and B. Ottersten, "Physical layer security in multibeam satellite systems," *IEEE Trans. Wireless Commun.*, vol. 11, no. 2, pp. 852–863, 2012.

53. Z. Lin, M. Lin, W.-P . Zhu, J.-B. Wang, and J. Cheng, "Robust secure beamforming for wireless powered cognitive satellite-terrestrial networks," *IEEE Trans. Cognitive Commun. Network.*, vol. 7, no. 2, pp. 567–580, 2021.
54. L. Lv, H. Jiang, Z. Ding, Q. Ye, N. Al-Dhahir and J. Chen, "Secure non-orthogonal multiple access: An interference engineering perspective," *IEEE Netw.*, vol. 35, no. 4, pp. 278–285, July/August 2021.
55. L. Lv H. Jiang, Z. Ding, L. Yang and J. Chen, "Secrecy-enhancing design for cooperative downlink and uplink NOMA with an untrusted relay," *IEEE Trans. Commun.*, vol. 68, no. 3, pp. 1698–1715, 2020.
56. K. An, M. Lin, W. Zhu, Y. Huang, and G. Zheng, "Outage performance of cognitive satellite terrestrial networks with interference constraint," *IEEE Trans. Vehi. Tech.*, vol. 65, no. 11, pp. 9397–9404, Jan. 2016.
57. Z. Lin, M. Lin, J. Wang, T. de Cola, and J. Wang, "Joint beamforming and power allocation for satellite-terrestrial integrated networks with non-orthogonal multiple access," *IEEE J. Sel. Top. Signal Process.*, vol. 13, no. 3, pp. 657–670, 2019.
58. H. Mei, K. Yang, Q. Liu and K. Wang, "3D-trajectory and phase-shift design for RIS-assisted UAV systems using deep reinforcement learning," *IEEE Trans. Veh. Technol.*, vol. 71, no. 3, pp. 3020–3029, 2022.
59. N. I. Miridakis, D. D. Vergados, and A. Michalas, "Dual-hop communication over a satellite relay and shadowed Rician channels," *IEEE Trans. Vehi. Tech.*, vol. 64, no. 9, pp. 4031–4040, Sep. 2015.

Elevating Connectivity: Integrating Satellite Access Networks in 3GPP NTN Vision

Malik Muhammad Saad, Muhammad Ashar Tariq, and Dongkyun Kim

Abstract

The rapid evolution of communication technologies has paved the way for a paradigm shift in network infrastructure, particularly in the context of Non-Terrestrial Networks (NTN). This article offers an in-depth exploration of the remarkable advancements catalyzed by the 3rd Generation Partnership Project (3GPP), notably with the advent of Releases 17 and 18. At the forefront of this transformation is the integration of satellite access networks into terrestrial counterparts, a breakthrough that demands a unified, multi-dimensional design approach. From the fundamental Radio Access Network (RAN) to intricate services and system aspects, extending to core infrastructure and terminal equipment, every facet undergoes profound scrutiny. The article meticulously examines the implications of satellite access network incorporation, both as a newfound 3GPP RAN and as a pivotal backhaul solution. Central to this analysis are the seismic shifts in network architecture and performance metrics, taking into account critical factors such as coverage, capacity, and latency. Delving into the crucible of 3GPP Release 18, the article identifies and addresses key challenges, proffering innovative solutions within the ambit of the NTN framework. These solutions are poised to be instrumental in fortifying the integrated

M. M. Saad · M. A. Tariq · D. Kim (✉)
School of Computer Science and Engineering, Kyungpook National University, Daegu, Korea
e-mail: dongkyun@knu.ac.kr

© The Author(s), under exclusive license to Springer Nature Switzerland AG 2025

M. A. Jamshed and A. Nauman (eds.), *Integrated Terrestrial and Non-Terrestrial Networks*, Synthesis Lectures on Communications,
https://doi.org/10.1007/978-3-031-63705-6_8

architecture necessary to propel 5G and beyond into the future. This comprehensive analysis serves as a critical compass, guiding stakeholders, technologists, and policymakers through the dynamic terrain of NTN, illuminating the path toward a more interconnected and technologically advanced era of global communications.

Keywords

Non-terrestrial networks (NTN) • Satellite radio access network • 3GPP new ran • Non-terrestrial access (NTA)

1 Introduction

In the ever-evolving landscape of communication technologies, the rise of Non-Terrestrial Networks (NTN) serves as a testament to the transformative potential of innovation. The Third Generation Partnership Project (3GPP) has been a driving influence in advancing this evolution, with Releases 17 and 18 standing as monumental milestones. At the core of this paradigm shift is the integration of satellite access networks seamlessly into terrestrial counterparts. This transformative convergence necessitates a comprehensive and multi-faceted design approach, addressing every aspect of the network infrastructure, from the fundamental Radio Access Network (RAN) to the complexities of services, system elements, core infrastructure, and terminal equipment [1, 2].

The incorporation of satellite access networks into the 3GPP ecosystem not only establishes a novel 3GPP RAN but also heralds a transformative backhaul solution. This chapter embarks on a meticulous examination of the far-reaching implications of this integration, dissecting the core facets of network architecture and performance metrics. Critical considerations such as coverage, capacity, and latency come under intense examination, as the seismic shifts brought forth by this integration reshape the landscape of global connectivity [3].

Within the dynamic realm of NTN, this comprehensive analysis serves as a critical compass. It guides stakeholders, technologists, and policymakers through the intricacies of integrated satellite access networks, offering insights that transcend the boundaries of traditional communication paradigms. The chapter positions itself as a cornerstone in navigating the dynamic landscape of NTN, providing a clear vision for the future of global communications.

In a world where connectivity is not merely a necessity but a fundamental enabler of progress, the integration of satellite access networks in 3GPP NTN vision assumes paramount importance. The interconnectedness of our digital age requires a robust and flexible infrastructure that can seamlessly adapt to the diverse needs of users and applications [4]. The convergence of satellite and terrestrial networks, as facilitated by 3GPP Releases 17 and 18, extends beyond the conventional boundaries of communication, fostering a symbiotic connection.

The infusion of satellite capabilities into the RAN not only extends coverage to previously underserved areas but also enhances the overall resilience and reliability of the network. As the analysis extends to the core infrastructure and terminal equipment, it becomes evident that the integration of satellite access networks is not merely a technical feat but a strategic imperative. The alignment of satellite capabilities with terrestrial networks in the backhaul architecture optimizes resource utilization and enhances overall network efficiency [5]. However, the integration journey is not without its challenges, and this article confronts them head-on. Drawing insights from the challenges identified in 3GPP Release 18 technical reports [6], this chapter delves into the discussion of these challenges and potential solutions. The exploration is grounded in the context of the 3GPP proposed work items and ongoing research. To this end, Sect. 2 presents the integration of satellite access networks into terrestrial networks. Section 3 delves into the challenges and solutions in the context of the impact of satellite next-generation RAN (NG-RAN) in 3GPP architecture for 5GC. Section 4 examines the impact of satellite NG-RAN on backhaul for 5GC. Section 5 explores the integration of satellite technology into terminal and core network functionalities, and Sect. 6 discuss from the RAN perspective within the 5G system (5GS). Section 7 presents potential solutions considering the proposed work items in 3GPP Release 18. Finally, conclusions are drawn in Sect. 8.

2 Integration of Satellite Access Networks Within the Realm of 3GPP Architecture

The integration of satellite access networks into terrestrial counterparts marks a transformative phase in communication technologies, with a focus on harmonizing the capabilities of satellite components within the existing terrestrial infrastructure. 3GPP aims to establish a comprehensive framework that facilitates robust communication, seamless transitions between networks, optimal allocation of resources, and effortless management of mobility, ensuring uninterrupted connectivity between the 5GC and the satellite-based NG-RAN [7]. This holistic approach spans across various dimensions, encompassing everything from the fundamental RAN to complex services, system elements, core infrastructure, and terminal equipment. By seamlessly integrating satellite access networks into the 3GPP ecosystem, this amalgamation ensures a unified and adaptable architecture capable of catering to the diverse requirements of users and applications.

The importance of a unified approach lies in its capacity to transcend traditional communication paradigms. It not only establishes a novel 3GPP RAN but also heralds a transformative backhaul solution. Figure 1 illustrates the satellite network within the 3GPP RAN architecture and its role as a backhaul.

3 Satellite NG-RAN in 3GPP Architecture

The integration of Satellite NG-RAN presents a novel approach within the 3GPP framework, allowing for seamless incorporation of satellite connectivity into the 5GC Network [2]. This integration involves ensuring uninterrupted and efficient communication channels between

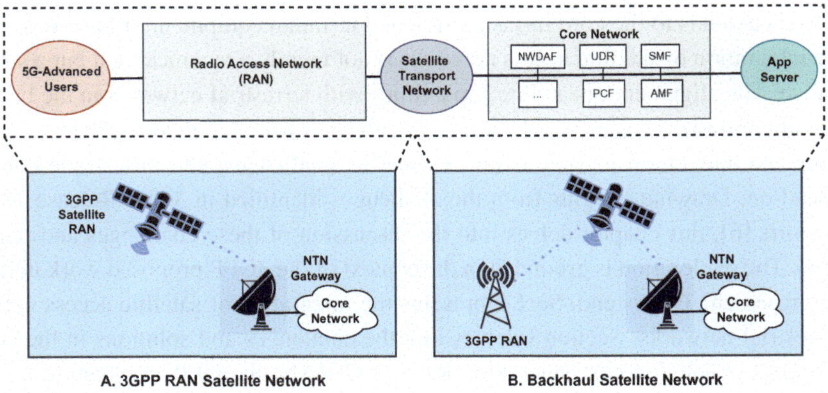

Fig. 1 Integration of the satellite network within the 3GPP RAN architecture and its role as a Backhaul [1]

User Equipment (UE) and satellite NG-RAN, while concurrently tackling challenges associated with maximizing resource utilization.

With the rise of Internet of Things (IoT) systems, accommodating a diverse range of devices in various access environments becomes increasingly important. This necessitates the utilization of diverse wireless spectrums to support the evolving requirements of IoT applications. Within the scope of 3GPP Release 17 & 18, there is notable attention placed on establishing direct access using transparent satellites. This initiative aims to streamline connectivity by bypassing intermediary nodes, enhancing efficiency and reducing latency within the 5GC ecosystem. The impact of this integration on the 5GC is intricately linked to the distinct characteristics of satellite cells, which often surpass the size of their terrestrial counterparts [2]. These satellite cells may maintain a stationary position on Earth with adjustable beams or traverse the Earth's surface with fixed beams, presenting unique challenges and opportunities.

As satellite cells exhibit differing characteristics from terrestrial cells, adjustments become imperative in various aspects of network management. Mobility management strategies need to be adapted to account for the dynamic nature of mobile satellite cells, ensuring seamless handovers and continuity of service. Similarly, handover processes must be optimized to accommodate the distinct movement patterns and coverage areas of satellite cells. Furthermore, ensuring User Equipment (UE) accessibility for downlink services becomes paramount, necessitating tailored approaches to maintain connectivity and optimize data transmission efficiency.

Addressing these challenges requires a nuanced understanding of satellite cell behavior and its integration within the 5GC architecture. By implementing tailored mobility management, handover optimization, and UE accessibility strategies, network operators can harness the full potential of satellite connectivity while ensuring seamless integration within the 5GC framework.

In the subsequent sections, we will delve into the challenges and potential solutions associated with the integration of satellite NG-RAN with the 3GPP 5GC, aiming to provide a comprehensive understanding of this innovative approach within the telecommunications landscape.

3.1 Impact of Satellite NG-RAN on 5GC Mobility Management

The integration of Satellite NG-RAN within the 3GPP architecture introduces significant implications for the mobility management of the 5GC [8]. This section explores the nuanced adjustments required to accommodate the larger cell size of satellite cells, emphasizing the role of handover procedures in ensuring uninterrupted connectivity across diverse network nodes.

3.1.1 Handover Adjustments in Satellite NG-RAN

The extensive cell size of Satellite NG-RAN requires meticulous adjustments in handover procedures. Addressing the unique characteristics of satellite beams, including their size and configurations (fixed or mobile cells), the focus is on achieving seamless connectivity and maintaining optimal communication between UE and Satellite NG-RAN [9].

3.1.2 Considerations for Network Node Connectivity

Apart from individual handovers, the integration of Satellite NG-RAN has implications for connectivity between different network nodes. This includes an examination of how Satellite NG-RAN influences the overall network architecture, with an emphasis on the need for cohesion and efficient data transfer across the 5GC. Challenges and solutions related to network node connectivity, considering regenerative payload and transparent payload, in the context of satellite NG-RAN are discussed in [10, 11].

3.2 Challenges in Management of UE Reachability with Satellite NG-RAN

Deploying Satellite NG-RAN presents unique hurdles in accurately pinpointing the locations of UE for downlink services [12]. This necessitates tackling the intricacies introduced by satellite beams-whether they are steerable or fixed, stationary or mobile-and implementing adjustments to ensure effective paging and seamless service provision within satellite coverage zones.

3.2.1 Adaptations for Efficient Paging

Efficient paging in the context of satellite coverage addresses challenges posed by varying beam characteristics. This involves implementing strategies and adaptations to streamline the paging process, ensuring prompt and reliable communication between the 5GC and UEs utilizing Satellite NG-RAN.

3.2.2 Uninterrupted Service Delivery

Ensuring continuous service delivery in the presence of satellite beam dynamics involves implementing measures to guarantee uninterrupted service availability for UEs within satellite coverage areas. Factors such as beam mobility and variability are considered in this context [13].

3.3 Management of Tracking Area Codes (TAC) with Mobile Satellite Cells

Deploying mobile satellite cells leads to variations in Tracking Area Codes (TAC), necessitating robust management approaches. To address this, 3GPP suggests broadcasting a list of TACs associated with specific geographical tracking areas. The primary focus is on ensuring accurate recognition and efficient supervision of UE locations within the coverage areas of satellite cells [14].

3.3.1 Ensuring Stability in UE Location Management

In the technical context, the broadcasting mechanism proposed by 3GPP for TAC outlines how this approach facilitates the identification of specific tracking areas defined by geographical considerations. This ensures stability and reliability in UE location management. Building upon the broadcasted TAC list, measures should be taken to guarantee stability in UE location management. This involves detailing how consistent TACs, even in the presence of moving satellite cells, contribute to the overall stability of the 5GC, providing reliable services to users.

4 Satellite Radio Access as a Backhaul Solution

When incorporating a satellite NG-RAN for backhaul purposes, it serves as a vital link between a Public Land Mobile Network (PLMN) and a satellite network, enabling seamless connectivity and data transmission [15]. Within the domain of backhauling, 3GPP Release 18 places significant emphasis on scenarios where the satellite NG-RAN plays a pivotal role. By leveraging satellite technology for backhaul purposes, 3GPP aims to address the challenges associated with extending network coverage to areas where traditional terrestrial

infrastructure may be impractical or cost-prohibitive. One of the primary objectives is to address and minimize consistent delays inherent in satellite communication, thus mitigating their impact on the overall network architecture.
In these scenarios, network operators adopt strategic measures to effectively manage and optimize the performance of the satellite backhaul. One such approach involves leveraging satellite ephemeris data, which provides precise information about the position and movement of satellites. By utilizing this data, operators can calculate variable delays, compensating for fluctuations in both the service and feeder links. This proactive measure ensures that the transmission of data between the satellite NG-RAN and the core network remains efficient and reliable, despite potential challenges posed by signal latency.
Furthermore, the choice of satellite orbits also plays a crucial role in optimizing backhaul performance. Operators may opt for either Geostationary Earth Orbit (GEO) satellites or Non-Geostationary Satellite Orbit (NGSO) satellites, depending on factors such as coverage area, bandwidth requirements, and latency considerations. Importantly, in scenarios where Inter-Satellite Links (ISLs) are not utilized, the focus is on maximizing the efficiency of the chosen satellite constellation to meet the demands of backhaul connectivity effectively [1].
Overall, integrating satellite NG-RAN into the backhaul infrastructure involves careful planning and optimization to ensure seamless communication between the PLMN and the satellite network. By addressing challenges related to latency and leveraging innovative solutions, operators can establish robust backhaul connections that enhance network performance and reliability, ultimately benefiting end-users with improved connectivity experiences.

5 Incorporating Satellites Into the 5GS: Aspects Regarding Terminals and Core Infrastructure

Satellite integration within the 5GS presents unique challenges and opportunities for terminal devices and the core network [16]. The following comprehensive discussion explores key aspects related to the terminal and core network, addressing critical issues and proposing solutions for the seamless integration of satellite communication into the 5G ecosystem.

5.1 PLMN Selection Procedure Enhancement

The PLMN selection procedure is a fundamental aspect of ensuring effective satellite integration. In response to the satellite integration in 5GS, the PLMN selection procedure has undergone updates [15]. Notable specifications include [1, 17]:

- Introduction of the "NG-RAN satellite" Radio Access Technology (RAT) type in User Identity Module (USIM).
- Extension of Non Access Stratum (NAS) supervision timers for satellite access.

- Modification of higher priority PLMN selection to include shared Mobile Country Code (MCC).
- Implementation of new triggers and cause values for PLMN selection based on user and network conditions.
- Support for multiple TACs for the same PLMN broadcast in the radio cell.

Following are the key issues and the potential solutions.

5.1.1 Determination of UE's Country Location

Addressing the challenge of determining the country of a UE physical location is critical. Solutions include leveraging Global Navigation Satellite System (GNSS) capabilities and implementing geolocation mapping within the UE. This technology enables the association of precise geolocation coordinates with specific countries.

5.1.2 Use of Core Network in UE's Country

Ensuring the enforcement of core network usage in the country where the UE is physically located involves considerations of legal and regulatory aspects. Solutions propose UE-configured lists of permissible PLMNs based on country and network-provided information.

5.1.3 PLMN Selection in International Areas

Given the global coverage of satellite access, handling PLMN selection in international areas is vital. Solutions include triggers for PLMN selection upon entering/exiting international areas and optional UE configurations for PLMN selection priorities in these regions.

5.1.4 Handling Shared/Global PLMN IDs

Shared or global PLMN IDs, represented by MCC, pose challenges in the legacy PLMN selection procedure. Solutions suggest modifications to accommodate these IDs, including new UE configurations and updates to steering of roaming functionality.

5.1.5 PLMN Search for Satellite Access

Defining the search interval for periodic PLMN search in satellite access is essential. Solutions explore the need for different priority levels, search intervals, and considerations for MCC matching in satellite access.

6 Satellite Network Integration in the 5GS: An Analysis from the RAN Aspect

Incorporating satellite technology into the 5GS from the RAN perspective represents a strategic evolution aimed at expanding connectivity options through the adoption of Non-Terrestrial Access (NTA). This approach allows for the establishment of network connections using satellites or High Altitude Platform Stations (HAPS) [5].

The NTA architecture comprises two key components: the NTN payload deployed on a satellite and the NTN Gateway positioned on the ground. These elements are interconnected through a feeder link, forming a cohesive network infrastructure. Within this setup, the NTN payload functions as an essential network node situated on the satellite itself. Its primary role is to facilitate communication and data transmission between the satellite and the ground-based components of the network. Figure 2 illustrates the architecture of NTA.

Conversely, the NTN Gateway serves as a critical terrestrial hub responsible for managing and coordinating communication activities between the satellite-based NTN payload and the rest of the 5GS network. This ground-based component acts as a central point for routing data traffic, ensuring seamless connectivity and efficient data exchange between the satellite and terrestrial network segments.

By leveraging the NTN architecture, the integration of satellite technology into the 5GS RAN enables enhanced coverage and connectivity capabilities, particularly in areas where traditional terrestrial infrastructure may be limited or unavailable. Additionally, the utilization of satellites and HAPS opens up opportunities for extending network reach to remote or underserved regions, contributing to the broader goal of bridging the digital divide and enabling universal access to high-speed connectivity.

Within the framework of NTA, three specific categories of service links are introduced [1]:

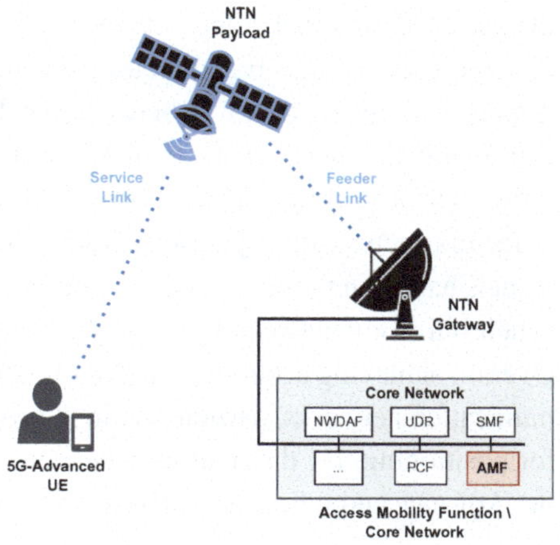

Fig. 2 Non terrestrial access architecture

- Earth-Fixed Service Link: Continuous coverage over specific geographical areas, applicable to GSO satellites.
- Quasi-Earth-Fixed Service Link: Coverage of a single region for a limited period, switching to cover a different area, commonly associated with NGSO satellites.
- This refers to the dynamic provision of coverage across various regions, facilitated by beams that dynamically adjust to cover different areas of the Earth's surface.

Incorporating satellite technology into the 5GS from the RAN perspective requires a range of enhancements and adjustments to current infrastructure and protocols [5, 18].

6.1 Achieving Precision in Time Coordination

In the process of synchronization, which is crucial for the seamless operation of communication networks, UEs connected to NTN must determine their precise positions using GNSS data. This synchronization process is essential for various network operations, including efficient handovers and resource allocation. To achieve this synchronization, the network distributes ephemeris data to UEs. Ephemeris data contains information about the position and movement of satellites in the GNSS constellation. This data is essential for UEs to accurately calculate their own positions relative to the satellites.

The distribution of ephemeris data can occur during different stages of UE operation. It may be provided during the initiation of access to the network or periodically while the UE is in connected mode. By continuously updating the ephemeris data, the network ensures that UEs maintain accurate positioning information, even as satellites move through space.

However, in addition to the challenge of determining accurate positions, there are also propagation delays that must be accounted for in the synchronization process. These delays occur as signals travel between the UE and the satellite. To compensate for these delays and ensure precise synchronization, a combination of techniques is employed. One such technique is the use of Common Timing Advance (CTA). CTA allows the network to adjust the timing of signals sent to UEs based on their distance from the satellite. By accounting for the time it takes for signals to travel between the UE and the satellite, CTA helps to synchronize communication between the two. Additionally, scheduling offsets may be applied to further refine synchronization. These offsets account for variations in signal propagation times due to factors such as atmospheric conditions or signal reflections. By fine-tuning the timing of signals based on these factors, scheduling offsets contribute to maintaining highly accurate synchronization between UEs and the NTN network [19].

Overall, achieving accurate time coordination is essential for the reliable operation of communication networks, particularly in the context of NTN where UEs rely on satellite signals for positioning. By distributing ephemeris data, employing techniques like Common Timing Advance and scheduling offsets, and continuously updating synchronization parameters,

NTN networks can ensure that UEs maintain precise positioning information, enabling efficient communication and network operation.

6.2 Advancements in Hybrid Automatic Repeat Request (HARQ)

This improvement focuses on refining the operation of HARQ mechanisms within the communication protocol stack. Specifically, it involves a strategic adjustment in how HARQ feedback is handled during Automatic Repeat Requests (ARQ) at the Radio Link Control (RLC) layer.

Traditionally, HARQ feedback mechanisms can sometimes introduce delays, leading to inefficiencies in data transmission. By selectively deactivating HARQ feedback during ARQ processes at the RLC layer, this enhancement aims to mitigate such delays, thereby reducing instances of HARQ stalling. This adjustment ensures that the communication flow remains smooth and uninterrupted, even in challenging network conditions.

Additionally, to further bolster reliability, there is an augmentation in the frequency of retransmissions at the Medium Access Control (MAC) layer. This means that if data packets are lost or corrupted during transmission, they are promptly retransmitted. By increasing the retransmission frequency, this measure enhances the overall robustness of the communication link, thereby improving the reliability of data delivery. Overall, these refinements in HARQ mechanisms contribute to enhancing the efficiency and reliability of data transmission in the communication network, ultimately leading to improved performance and user experience.

6.3 Managing User Mobility

Efficient user mobility management encompasses seamless handovers between the satellite access network and terrestrial access network. Roaming agreements are established to facilitate smooth transitions between NTN and terrestrial networks. These handovers are influenced by factors such as location, time, and specific events. The framework also incorporates triggering conditions to enable conditional handovers. This allows for transitions based on predetermined criteria related to location, time, and specific events. The aim is to optimize the user experience by ensuring connectivity transitions are seamless and contextually aligned.

6.4 Access Mobility Function (AMF) Re-selection

The AMF re-selection process plays a crucial role in maintaining seamless connectivity for UE within the network. Essentially, it ensures that UE remains connected to the most suit-

able AMF based on its current location relative to the serving cell. When a UE moves within the network coverage area, the gNB (next-generation NodeB or base station) continually monitors its location. If the gNB determines that the UE's current connection to the AMF is no longer optimal for efficient communication, it triggers the AMF re-selection process. This process involves identifying a new AMF that can better serve the UE's current location. The primary objective of AMF re-selection is to facilitate a smooth handover process, ensuring uninterrupted service continuity for the UE. By seamlessly transitioning the UE's connection to the new AMF, the network optimizes the UE's network performance and ensures a consistent quality of service as it moves across different geographical areas.
AMF re-selection is essential for optimizing UE connectivity within the network, allowing for efficient handovers and ensuring a seamless user experience, even as the UE moves within the network coverage area.

6.5 RAN Aspects for Non-Terrestrial Networks (NTN) in NR Protocol

Crucial aspects of the RAN for NTN within the New Radio (NR) protocol include the integration of NTN payload and NTN Gateway to enable non-terrestrial access. The NTN architecture supports three types of service links: Earth-Fixed, Quasi-Earth-Fixed, and Earth-Moving [17]. Additionally, RAN aspects for NTN cover timing synchronization, HARQ enhancements, mobility management, and AMF re-selection. This collectively establishes a robust framework for NTN operations [5, 18].

6.6 Positioning Enhancements

Positioning enhancements in the NR protocol focus on improving the accuracy, latency, and integrity of GNSS data. Solutions include addressing timing delay errors, enhancing uplink Angle of Arrival (UL-AOA) and downlink Angle of Departure (DL-AOD), and mitigating multi-path and Non-Line-of-Sight (NLOS) conditions. These measures collectively contribute to elevating the precision and reliability of positioning information within the 5GS.

6.7 RAN Slicing for NR

The introduction of the Network Slice Access stratum Group (NSAG) enables slice-aware cell reselection and configuration of Radio Access Channel (RACH) parameters specific to each network slice [20]. This feature supports service continuity, enforces Slice Maximum Bit Rate (MBR), and utilizes Target NSSAI (Network Slice Selection Assistance Information), contributing to the efficient implementation of RAN slicing within the NR framework.

6.8 On-Demand Positioning Reference Signal (PRS) Transmission

This procedure empowers the Location Management Function (LMF) to dynamically control PRS transmission characteristics for positioning measurements. By enabling on-demand adjustments, the system can optimize the transmission parameters based on the specific requirements of positioning measurements, enhancing flexibility and efficiency.

6.9 Positioning in RRC INACTIVE State

This feature extends support for NR positioning measurements and methods even when the UE is in the RRC inactive state. This ensures that positioning capabilities remain available even when the UE is not actively communicating with the network, enhancing the overall flexibility and responsiveness of the positioning system.

6.10 GNSS Integrity

The introduction of UE-based GNSS integrity establishes a trust mechanism for the accuracy of position-related data. This enhancement ensures that the UE can independently verify the integrity of GNSS signals, enhancing reliability and security in location-based services.

7 Integrating Satellite Components in 5G Networks: Potential Solutions

The integration of satellite components into 5G networks is a complex task that demands innovative solutions to ensure a seamless fusion of terrestrial and satellite technologies. Addressing challenges in network slice management, component coordination, and monitoring is crucial for achieving optimal performance [10]. Here, we delve into detailed solutions for each of these key areas.

7.1 Network Slice Management

In order to seamlessly integrate satellite components into 5G networks, an extension of the existing service profile [21] is proposed. This extension should accommodate the unique attributes and requirements of satellite RAN, facilitating the creation and management of a dedicated network slice instance associated with satellite components. The following are the two methods.

- Attribute Inclusion: Extend the service profile to include specific attributes related to satellite RAN, considering factors such as latency, coverage area, and handover procedures.
- Dynamic Management: Ensure adaptability to dynamic environmental factors, such as satellite movement and changing link conditions, by incorporating real-time updates into the service profile

7.2 Component Management

Efficient component management is essential for harmonizing the diverse technologies of satellite and terrestrial RANs. Extending the existing slice profile to specify separate service requirements for both satellite RAN and terrestrial RAN provides a solution for creating network slice subnet instances (NSSIs) with distinct characteristics. The following are the two methods.

- Performance Requirement Attribute: Introduce a performance requirement attribute within slice profile, distinguishing between satellite and terrestrial RAN performance criteria.
- Sevice Customization: Allow for the customization of service requirements in slice profile to cater to the specific needs and capabilities of both satellite and terrestrial components.

7.3 Adaptive DL-Air Interface Measurements for HARQ Feedback

Monitoring and feedback mechanisms are critical for ensuring the reliability of satellite links. Adapting the downlink (DL) air interface measurements (TS 28.552) to support scenarios where HARQ feedback is disabled for MEO and GEO satellite links is essential for robust satellite component monitoring. Following potential solutions are under consideration.

- New Feedback Mechanism: Integrate a new feedback mechanism that supports DL transmission feedback when HARQ feedback is disabled for satellite links.
- Standard Compliance: Update measurement conditions clauses in TS 28.552 to ensure compliance with the new feedback mechanism, accommodating both terrestrial and satellite communication scenarios.

7.4 Multi-RAT Load-Balancing for Both Satellite and Terrestrial RANs

Load-balancing between satellite and terrestrial components is crucial for optimal resource utilization [22]. Enabling the multi-Radio Access Technology (RAT) resource control module to make load-balancing decisions by switching traffic between satellite and terrestrial gNBs based on performance reports is a strategic approach. An alternative approach to load-balancing involves splitting traffic between satellite and terrestrial gNBs based on performance reports. Following are approaches to achieve load balancing.

- Resource Utilization Analytics: Leverage resource utilization analytics functions specified in TR28.809 to inform load-balancing decisions or to split traffic.
- Link Performance Monitoring: Implement robust monitoring of link performance between User Equipment (UEs), terrestrial gNBs, and core networks for accurate load-balancing decisions.

8 Conclusions

The article explored the complex integration of Non-Terrestrial Networks (NTN) into terrestrial networks, highlighting the advancements facilitated by the 3rd Generation Partnership Project (3GPP) in Release 17 and the proposed work plan for Release 18. This chapter serves as a comprehensive guide, offering insights into the technological advancements and their impacts within the 3GPP framework. It delves into the influence of satellite access networks on NTN, analyzing their dual role as both a novel 3GPP Radio Access Network (RAN) and a solution for backhaul connectivity. The analysis uncovered ongoing efforts by 3GPP RAN and CT groups to standardize the integration of NTN with terrestrial networks, with a focus on improving coverage and meeting evolving connectivity demands. The incorporation of satellite capabilities not only extends coverage to previously underserved areas but also enhances the network's reliability and resilience.

Acknowledgements This research has been supported by the National Research Foundation of Korea (NRF) grant funded by the Korea government (MSIT), (NRF-2022R1A2C1003620). In addition, this research was supported by the 2021 Kyungpook National University BK21 FOUR Graduate Innovation Project (International Joint Research Project for Graduate Students).

References

1. M. M. Saad, M. A. Tariq, M. T. R. Khan, and D. Kim, "Non-terrestrial networks: An overview of 3gpp release 17 & 18," *IEEE Internet of Things Magazine*, vol. 7, no. 1, pp. 20–26, 2024.
2. M. Harounabadi and T. Heyn, "Toward integration of 6g-ntn to terrestrial mobile networks: Research and standardization aspects," *IEEE Wireless Communications*, vol. 30, no. 6, pp. 20–26, 2023.
3. 3GPP, "Study on architecture aspects for using satellite access in 5G," Tech. Rep. TR 23.737, 2023.

4. X. Lin, S. Rommer, S. Euler, E. A. Yavuz, and R. S. Karlsson, "5g from space: An overview of 3gpp non-terrestrial networks," *IEEE Communications Standards Magazine*, vol. 5, no. 4, pp. 147–153, 2021.
5. 3GPP, "Technical Specification Group Services and System Aspects; Release 17 Description," Tech. Rep. TR 21.917 V17.0.1, 2023-01.
6. 3GPP, "3rd Generation Partnership Project; Technical Specification Group Radio Access Network; Non-Terrestrial Networks (NTN) L-/S-band for NR; (Release 18)," Tech. Rep. TR 38.741 V18.0.0, 2023-12.
7. G. Masini, P. Reininger, M. El Jaafari, A. Vesely, N. Chuberre, B. Baudry, and J.-M. Houssin, "5g meets satellite: Non-terrestrial network architecture and 3gpp," *International Journal of Satellite Communications and Networking*, vol. 41, no. 3, pp. 249–261, 2023.
8. F. Rinaldi, H.-L. Maattanen, J. Torsner, S. Pizzi, S. Andreev, A. Iera, Y. Koucheryavy, and G. Araniti, "Non-terrestrial networks in 5g & beyond: A survey," *IEEE access*, vol. 8, pp. 165178–165200, 2020.
9. E. Juan, M. Lauridsen, J. Wigard, and P. Mogensen, "Handover solutions for 5g low-earth orbit satellite networks," *IEEE Access*, vol. 10, pp. 93309–93325, 2022.
10. 3GPP, "Solutions for NR to support non-terrestrial networks (NTN)," Tech. Rep. TR 38.821, 2023.
11. S. Euler, X. Fu, S. Hellsten, C. Kefeder, O. Liberg, E. Medeiros, E. Nordell, D. Singh, P. Synnergren, E. Trojer, and I. Xirouchakis, "Using 3gpp technology for satellite communication," *Ericsson Technology Review*, vol. 2023, no. 6, pp. 2–12, 2023.
12. T. Norp, J.-Y. Fine, and R. Djapic, "Network aspects of integrating 5g and satellite communication," *International Journal of Satellite Communications and Networking*, 2023.
13. F. Rinaldi, H.-L. Määttänen, J. Torsner, S. Pizzi, S. Andreev, A. Iera, Y. Koucheryavy, and G. Araniti, "Broadcasting services over 5g nr enabled multi-beam non-terrestrial networks," *IEEE Transactions on Broadcasting*, vol. 67, no. 1, pp. 33–45, 2021.
14. 3GPP, "Procedures for the 5G System (5GS), Stage 2," Tech. Rep. TR 23.502, 2023.
15. 3GPP, "Study on PLMN selection for satellite access in 5G," Tech. Rep. TR 23.502, 2023.
16. 3GPP, "Service Requirements for the 5G System," Tech. Rep. TR 22.261, 2022.
17. G. M. A. G. M. E. J. Alessandro Vanelli-Coralli, Nicolas Chuberre, *5G Non-Terrestrial Networks: Technologies, Standards, and System Design*. Wiley, 2024.
18. 3GPP, "NR; User Equipment (UE) radio transmission and reception, part 5: Satellite access Radio Frequency (RF) and performance requirements (RAN4)," Tech. Rep. TS 38.101-5, 2023.
19. 3GPP, "NR; Satellite Access Node radio transmission and reception (RAN4)," Tech. Rep. TS 38.108, 2023.
20. 3GPP, "3rd Generation Partnership Project; Technical Specification Group Core Network and Terminals; Study on PLMN selection for satellite access (Release 17)," Tech. Rep. TR 24.821 V17.0.0, 2021.
21. 3GPP, "5G Management and orchestration; 5G Network Resource Model (NRM), Stage 2 and Stage 3," Tech. Rep. TS 28.541, 2023.
22. 3GPP, "Technical Specification Group Services and System Aspects Study on management and orchestration aspects of integrated satellite components in a 5G network (Release 17)," Tech. Rep. TR 28.808, 2021.

SPRINGER NATURE

GPSR Compliance

The European Union's (EU) General Product Safety Regulation (GPSR) is a set of rules that requires consumer products to be safe and our obligations to ensure this.

If you have any concerns about our products, you can contact us on ProductSafety@springernature.com

In case Publisher is established outside the EU, the EU authorized representative is:

Springer Nature Customer Service Center GmbH
Europaplatz 3
69115 Heidelberg, Germany

The manufacturer's authorised representative in the EU is Springer
Nature Customer Service Centre GmbH, Europaplatz 3, 69115 Heidelberg,
Germany. If you have any concerns regarding our products, please
contact ProductSafety@springernature.com

Printed and bound by CPI Group (UK) Ltd, Croydon, CR0 4YY
30/09/2026
02225584-0019